£12.95

Toc

Geography and energy: commercial energy systems and national policies

THEMES IN RESOURCE MANAGEMENT
Edited by Professor Bruce Mitchell, University of Waterloo

Already published

John Blunden: Mineral resources and their management
J. D. Chapman: **Geography and energy: commercial energy systems and
    national policies**
Paul F. J. Eagles: The planning and management of environmentally sensitive
    areas
R. L. Heathcote: Arid lands; their use and abuse
Adrian McDonald and David Kay: Water resources: issues and strategies
Francis Sandbach: Principles of pollution control
Stephen Smith: Recreation geography

Forthcoming title

John Pierce: Food, land and man

J. D. CHAPMAN

Geography and energy: commercial energy systems and national policies

Longman
Scientific &
Technical

Copublished in the United States with
John Wiley & Sons, Inc., New York

Longman Scientific & Technical,
Longman Group UK Limited,
Longman House, Burnt Mill, Harlow,
Essex CM20 2JE, England
and Associated Companies throughout the world

Copublished in the United States with
John Wiley & Sons, Inc., 605 Third Avenue, New York, NY 10158

First published 1989

British Library Cataloguing in Publication Data
Chapman, J. D.
    Geography and energy: commercial energy systems and
    national policies. – (Themes in resource management).
    1. Energy resources. Geographical aspects
    I. Title   II. Series
    333.79

    ISBN 0-582-30085-1

Library of Congress Cataloging-in-Publication Data
Chapman, J. D. (John Doneric), 1923–
    Geography and energy: commercial energy systems and
    national policies / J. D. Chapman.
       p.   cm. – (Themes in resource management)
    Bibliography: p.
    Includes index.
    ISBN 0-470-21188-1 (Wiley, USA only).
    1. Energy industries.   2. Geography, Economic.   3. Energy policy.
    I. Title.   II. Series.
    HD9502.A2C455 1989
    333.79 – dc 19

Phototypeset in Times New Roman
by Paston Press, Loddon, Norfolk

Produced by Longman Group (FE) Limited
Printed in Hong Kong

Contents

Contents

List of figures

List of figures

List of tables

Acknowledgements

We are grateful to the following for permission to reproduce copyright material:

Energy, Mines and Resources, Ottawa for fig. 2.8 (Canada EMR 1981); Energy Technology Division, Oxford for figs. 2.6 & 2.7 from pp. 2 & 6 (Bush and Chadwick 1979); McGraw-Hill Book Co. for fig. 2.2 from fig. 4.3, p. 54 (Fowler 1984); National Research Council of Canada and the author, E. P. Cockshutt for fig. 2.9 (Cockshutt 1973); New Zealand Energy Research and Development Committee for figs. 2.4 & 2.5 from pp. 14–15 (Beca *et al.* 1979); Nichols Publishing Co. for tables IIB & III from tables 1–5 (Appendix 2), pp. 251–254 (Crabbe & McBride 1978).

Whilst every effort has been made to trace the owners of copyright, in some cases this has proved impossible and we take this opportunity to offer our apologies to any authors whose rights may have been unwittingly infringed.

Foreword

The Themes in Resource Management series has several objectives. One is to identify and to examine substantive and enduring resource management and development problems. Attention will range from local to international scales, from developed to developing nations, from the public to the private sector, and from biophysical to political considerations.

A second objective is to assess responses to these management and development problems in a variety of world regions. Several responses are of particular interest but especially *research* and *action programmes*. The former involves the different types of analysis which have been generated by natural resource problems. The series will assess the kinds of problems being defined by investigators, the nature and adequacy of evidence being assembled, the kinds of interpretations and arguments being presented, the contributions to improving theoretical undestanding as well as resolving pressing problems, and the areas in which progress and frustration ae being experienced. The latter response involves the policies, programmes and projects being conceived and implemented to tackle complex and difficult problems. The series is concerned with reviewing their adequacy and effectiveness.

A third objective is to explore the way in which resource analysis, management and development might be made more complementary to one another. Too often analysts and managers go their separate ways. A good part of the blame for this situation must lie with the analysts who too frequently ignore or neglect the concerns of managers, unduly emphasize method and technique, and exclude explicit consideration of the managerial implications of their research. It is hoped that this series will demonstrate that research and analysis can contribute both to the development of theory and to the resolution of important societal problems.

John Chapman's book is the seventh in the Themes in Resource Management series. The primary focus is at the global scale, providing a formidable challenge regarding data. Various governments generate information about energy consumption and supply in different ways, and often information is not complete or reliable. Chapman drew upon a wide range of sources for information used in this book, and was particularly successful in obtaining the cooperation of some of the major private companies involved in the field of energy.

The book presents concepts and information about both energy consumption and supply, with particular emphasis upon commercial energy systems. After having presented basic information about the geography of consumption and supply, he goes on to consider the issues involved in projecting future patterns of supply and demand. The complex mix of variables affecting future patterns – prices, technology, wars, economic growth or recession – emphasizes the difficulty in projecting or anticipating what the future opportunities and needs will be.

Having provided information about the nature and structure of commercial energy systems, Chapman then addresses policy issues in the final two chapters. The first examines national energy policies in the Soviet Union, United States and Japan, and also reviews international organizations such as OPEC, IEA and CMEA. The second chapter provides a more detailed examination of energy policy in Canada, with emphasis upon the fossil fuels (coal, oil and natural gas).

John Chapman has been an observer and analyst of the energy scene for three decades. His studies in energy began long before this field became 'popularized' as a result of the sharp increase of petroleum prices in the mid 1970s. As a result, this book has been written by an individual who has a particularly well developed appreciation of energy and energy systems. I think it is only such a person who could have woven together the disparate pieces of evidence in such a complete manner as has been done here.

Bruce Mitchell
University of Waterloo
Waterloo, Ontario

February 1988

Preface

This book grows out of twenty years of teaching a senior undergraduate course on the Geography of Commercial Energy. Over the years the course has interested not only students in geography but others from economics, political science, commerce and business administration as well as some from science and applied science disciplines. All had an interest in energy matters and, often, some familiarity with the energy industry from the point of view of their own discipline. However, few had sufficient background knowledge to permit a comprehensive understanding of the complexities of the geography of commercial energy systems or the national or international policies which attempt to manage them.

This book is intended to contribute to the development of an understanding of the working of energy systems, to a knowledge of their spatial structure, and to an introduction to the components of public policy and management. Spatially and temporally the empirical information is presented at a relatively high level of generalization, dealing mainly with major international economic groups and individual countries over decades rather than intranational divisions or particular siting problems. The conceptual frameworks that are reviewed are also broad to match these general spatial and temporal scales as well as to provide a reasonably comprehensive basis from which to approach more immediate and localized energy issues. The functional and other structural components of both the consumption and supply side of commercial energy systems are presented before the spatial aspects are dealt with. The focus is upon large scale commercial energy systems rather than traditional or emerging systems which deserve separate treatment in another volume. Chapter 1 introduces some basic concepts of energy systems and a broad interpretation of energetics. This is followed by two chapters on each of energy consumption and supply. Chapter 6 provides an overview of energy futures and the concluding chapters deal with energy policy at the international and national scale respectively.

CHAPTER 1

Energy and energetics

> Only those men who can convert heat and other forms of energy to work
> and can apply that work where they will, can travel over the world and shape
> it to their ends. The crux of the matter is the generation of work – the
> conversion of energy and its delivery to the point of application. (Luten
> 1971, p. 165.)

For the majority of people every day is devoted to work, often hard physical
work which leaves them drained. The energy they use is their own, fuelled by
an intake of food, and that of their animals whose energy is sustained by feed.
For others, such as the readers of this book, work means use of the mind in a
learning, critical way which involves little or no physical activity. While the laws
and regularities of scientific energetics underlie the human use of energy and
the work it can perform (Fowler 1975; Slesser 1978), from the social science
point of view it is also useful to consider the energy/work relationship in a
broader, less rigorous context. In these terms work may be thought of as the
effort exerted to achieve a chosen objective. All such effort is initiated by
human needs and motivations, involves some degree of intelligent guidance
and judgement, and ranges from largely physical to largely mental.

Types of work

Figure 1.1 portrays a broad typology of the work undertaken by humans with
four major types in each of the physical and mental categories. Much physical
work is directed towards modification of the environment or is aimed at
procuring and processing resources derived from the environment. The remainder arises from the transportation of goods and people. Mental work essentially
deals with the innovation, processing, application and communication of
information.

1

Physical	Modification of environment
	Clearing, sculpting and tilling land
	Modifying hydrologic regimes
	Constructing and conditioning built space
	Appropriating resources
	Harvesting and extracting
	Dumping wastes
	Processing material
	Concentrating, refining and synthesizing
	Assembling and re-cycling
	Transfer
	Transportation of material and people
Mental	Innovating information
	Technological research and development
	Creative arts
	Processing information
	Learning, recording, storing, manipulating
	Applying information
	Management and professional services
	Communicating information
	Teaching, media

Figure 1.1 A typology of work

Physical

The major tasks undertaken by mankind in the *modification of the environment* have been clearing land of existing vegetation in preparation for agricultural (or other) use, the impoundment and re-routeing of rivers, the reclamation of wetland and the construction of a built environment. With some notable regional exceptions (e.g. the Amazon River basin), land clearing has been completed but land sculpting is a continuing and growing type of work associated with engineering projects of all kinds. The construction of buildings and the associated conditioning of the enclosed space to provide an amenable temperature and lighting environment, is also a continuing activity in all societies.

The *appropriation of resources* involves activities ranging from primitive collecting and gathering to large scale machine harvesting of biomass (trees, crops) and extraction of minerals. In some societies, handling waste occasions large amounts of work while in others it is disposed of with as little effort as possible into the nearest available biophysical receptor.

Processing resources and materials is another major type of work in all societies. In less developed economies hundreds of millions of people are involved in the cooking and preservation of food, the simple processing of materials, and the making of clothing. In developed economies, the household tasks are augmented by the large-scale and complex processing of raw materials and the manufacture of a multitude of consumer and durable products.

All human groups expend effort on *transferring* themselves and materials from place to place. The journey to work and back, travel to central places to acquire or deliver goods and services or for social interaction are some of the transportation functions common to all societies. In recent decades the amount of work expended to accomplish spatial interaction has increased rapidly.

Mental

One of the distinguishing characteristics of post-industrial urban societies is the extent to which the labour force engages in mental rather than physical work. In many economies a major portion of the employment is in the quaternary sector and daily work involves research and development, word and data processing, marketing and the provision of specialized services to individuals and firms. The shift from physical to mental work has important implications both for the amount of and type of energy required (Tribus and McIrvine 1971).

Sources of energy

How much of the various types of work identified above can be accomplished and at what rate depends upon the sources of energy that are available. Primary energy sources commonly are grouped into the renewable and non-renewable classes discussed below. Recognition should also be given to the animate–inanimate dichotomy because it focuses attention upon the distinction between primary energy being converted into useful energy by organisms and that which is converted by machines and appliances (the 'slave' and 'robot' societies described in Zimmermann 1950).

Renewable and non-renewable

Figure 1.2 shows the primary sources of energy grouped into the two conventional categories of renewable and non-renewable. The concept of renewability is based on the time taken to replace the supply in relation to the time-scale of human events. Of the three renewable sources shown, gravity is persistent and ubiquitous but remains only a potential energy source unless associated with the motion of considerable mass (e.g. river flow, tides). Nuclear energy, in the form of geothermal heat, is also persistent and ubiquitous at great depths beneath the surface of the earth. Shallow or surface manifestations, however, are limited in distribution and, to the extent that the heat energy is carried by water (or steam), their renewability is subject to a continuous supply of water. Solar supply systems are based upon the continuous emission of radiation from the sun which reaches the surface of the earth discontinuously (diurnal and seasonal variations) and, as the result of cloud cover, somewhat unpredictably.

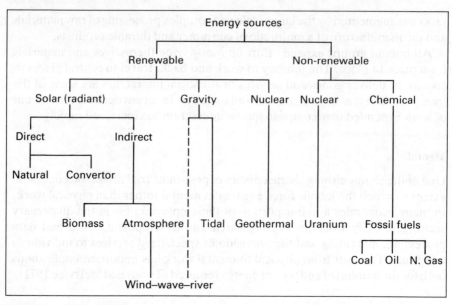

Figure 1.2 Energy sources

Consequently, all solar based energy supply systems, although renewable, are intermittent and, to some extent, cyclic. Furthermore, most of them are 'low-density' in the sense that the energy available per square metre is relatively small.

The use of *biomass* energy sources is characteristic of the rural, developing world in which millions of people and their draft animals carry out the bulk of subsistence work. Their muscle power (metabolic energy) is augmented by heat from burning firewood, crop and animal wastes in primitive convertors as well as from direct sunlight used for drying and preserving agricultural products. Such sources are renewable provided the rate of use and productive capacity of the biomass are kept in balance. Despite the large number of people involved, energy densities are low and the spatial range of any particular source is limited. Individual elements of biomass energy chains usually have low capital costs, are user-oriented and organized on a community basis.

Non-renewable energy sources are those which cannot be replenished within the span of human time. They represent the highly concentrated storage of massive amounts of material accumulated over millions of years of geological activity. However, despite the fact that these sources are non-renewable, so long as individual stocks remain and political stability prevails, a continuous and regular flow of energy can be supplied from them. Fossil fuels are subdivided into solid (coal, lignite) and liquid fuels (oil products), natural gas and nuclear fuels. Although each of these has a high energy content per unit of weight or volume in comparison with unprocessed biomass sources, they differ from one another in a number of important respects. The mining of solid fuels is relatively labour intensive and requires an elaborate and costly infrastructure, they can only be transported over long distances in batch containers (e.g. rail cars or vessels) and, when burned, frequently produce large quantities of gaseous and particulate residuals. The primary production facilities required for self-flowing oil and natural gas, on the other hand, need little labour input and, once brought to the surface, they may be transported continuously (pipelines) or in batches. Both, however, have to be more or less elaborately processed before use in capital-intensive oil-refineries and natural gas processing plants. Nuclear fuels have the highest energy content per unit of weight but involve the most elaborate, technically complex and expensive facilities, can only be used to generate electricity and require extreme care in handling.

Choosing among sources

The sources of energy selected for use by different societies vary in response to a number of factors which may be grouped into the following categories:

1. Occurrence – Many energy sources are confined to specific environments and locations and are thus only available at other locations when transport systems exist. Even physically present sources may not actually be available because of technical, economic or other constraints.
2. Transferability – The distance over which an energy source may be transported is a function of its physical form, energy content and transport technology.
3. Energy content – The amount of usable energy by weight or volume of a given source. Low energy content sources are inadequate when demand is large and spatially concentrated.
4. Reliability – Uninterrupted availability gives one source an advantage over another which is intermittent.
5. Storability – To meet interruptions of supply or peaks of demand, a source which can be stored has an advantage over one which cannot.
6. Flexibility – The greater the variety of end uses to which a given source or form may be put the more desirable it is.
7. Safety and impact – Sources which may be produced or used with low risk to human health and the environment will be preferred over less benign sources.
8. Cleanliness and convenience – The cleaner and more convenient source will be preferred over the dirty and cumbersome.
9. Price – The lower price source or form will be preferred over the more expensive.

In response to these factors, each energy using group selects a particular mix of energy sources to meet its needs. Such mixes change with time and vary from place to place. On a global scale the major sources of primary energy have changed from biomass and muscular energy, to coal, to oil products and natural gas and, most recently, to nuclear fuels. In terms of Figure 1.3 some economies

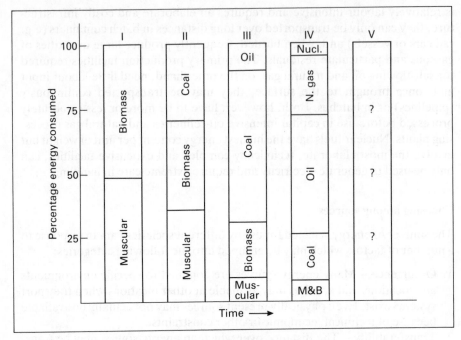

Figure 1.3 The evolution of energy consumption mixes by source

have moved from phase I to phase IV rapidly (e.g. Japan and the USSR) and others, particularly in the developing world, are having difficulty in moving beyond phase I in other than their major urban centres.

Energy systems

In order to mobilize and deliver energy from the primary energy sources identified in the previous section, all societies must create more or less elaborate energy supply systems. As a basis for the description of these systems it is helpful to consider first their general structure and, in order to understand the characteristics of particular systems, to identify the major variables which influence them (Mitsch *et al.* 1982).

Structure

Consideration of the basic structure of energy systems starts with the identification of the sources of energy in use and the whole array of physical facilities that have to be in place to link the chain of functions from initial production to end use. While these techno-physical facilities constitute the *functional* core of energy systems they are only one part of the total structure. The very large capitalization and cash flows involved and the resulting size of controlling

6

Structure	Variables
1. Functional	1. Technology
2. Economic	2. Benefits and costs
3. Organizational	3. Regulation
4. Spatial	4. Values and perceptions
5. Environmental	5. Resources and residuals
6. Social	6. Distance and density

Figure 1.4 The structural components and determining variables of energy systems

enterprises draw attention to the *economic* and *organizational* components. Beyond these are the *spatial*, *environmental* and *social* components which arise from the varied areal distribution patterns and densities of energy facilities and their attendant interrelations with biophysical and social systems (Fig. 1.4). In effect, energy systems have a multi-component structure within which each component has some influence upon the character of the others.

Functional   Many of the physical manifestations of energy systems are familiar to us all but it is not so obvious that each of them is a link in an energy chain which connects, often over great distances, a sequence of facilities extending from the production of primary energy to its end use. The general form of such chains is shown in Figure 1.5.

Stage I involves exploration for and the delineation, inventory and assessment of primary energy sources. For fossil and nuclear fuels, these activities require increasingly complex and costly operations particularly as they spread into inaccessible and difficult bio-physical environments. For most renewable resources the exploration/discovery phases are basically complete but much inventory and assessment has yet to be carried out. The latter frequently involves the establishment of instrumental monitoring networks to determine, for example, precipitation and streamflow or the incidence of solar radiation. Usually such networks need to be in operation for a decade or more before sufficient data are generated to permit evaluation.

Stage II, the production of primary energy, includes operations and equipment as different as individuals with a machete gathering firewood, a large open pit coal mine with its immense equipment and the relatively unobtrusive 'Christmas trees' of a producing natural gas field. Stage III, processing and transforming the primary energy into secondary energy, again may be as simple and inexpensive as a charcoal burner or biogas convertor or as complex and capital consuming as an oil refinery or thermal electric station. The end use of energy, Stage IV, is effected by a multiplicity of different types of convertors. Some can be made or assembled with little capital outlay but the great majority are specialized and require investment varying from tens to millions of dollars.

7

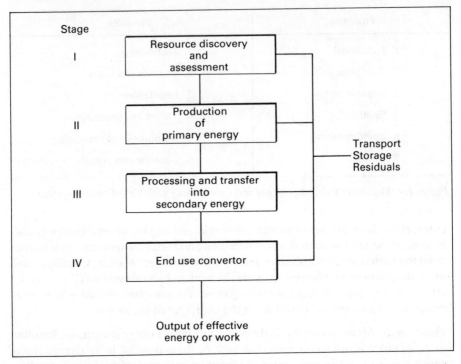

Figure 1.5 The general form of energy chains

Transportation facilities are required in and between most stages of the energy chain. In the first stage elaborate and special facilities are often required to assemble and operate the exploration and assessment equipment. Once production occurs, large capacity carrier systems are needed for the trunk transportation of, for example, crude oil or coal and distribution systems from the processing and transformation nodes reach out to innumerable individual consumers.

Storage facilities may be required in stages II-IV or between them. Some energy sources and forms are easily storable (e.g. solid fuels in stockpiles or liquid fuels in tanks) while for others, such as solar radiation and wind, storage of the primary energy is not an option. Similarly, beyond the limited storage provided by batteries, electricity cannot be stockpiled though of course the means of generating it can (e.g. fossil fuels and water behind a dam).

The production of residuals also takes place at all stages: exploration camp detritus, mine waste and dust at coal fields, sulphur and heat at processing and transformation facilities and various chemicals and particulates associated with the end use combustion of fuels. In order to dispose of these residuals, separate and large scale facilities are increasingly required.

Economic   The economic component of energy systems is frequently considered at two scales corresponding to the micro and macro approaches of economists. The former addresses the financial aspects of the energy firms in terms of such measures as capital assets and debt structure, gross expenditures and revenues, and returns on investment. Beyond the financial considerations, however, are the relations between the energy industry and economic activity generally. At this macro-scale attention is focused upon such broad issues as the economics of the selection of primary energy sources for development, the market for energy and the complexities of costs and prices, trade balances and currency stability.

Organizational The sheer functional and financial scope of commercial energy enterprises leads to large and complex organizations, both operational and regulatory. The operational entities differ in terms of ownership and control, degree of vertical integration within the energy chain, horizontal integration between energy sources and the spatial extent of their operations. Some organizations, epitomized by the multi-national energy companies, have exerted economic and political influence to a degree that challenges the power of governments. The formal, hierarchically structured public and private corporations of the developed world stand in marked contrast to the individual user, community and small entrepreneurial structure of the rural, biomass based energy systems of the developing world.

Spatial The spatial component of energy systems involves: consideration of the location of the stages of individual energy chains and the patterns and densities which result, the spatial association of these locations with elements of the biophysical environment and the social fabric of the areas concerned, the spatial interactions resulting from the separation of production and consumption, and the evolution of characteristic regional energy systems. In most fossil fuel and hydro-electric energy chains the tendency has been for ever larger production facilities with corresponding spatial concentration of energy supply. At the same time, the very large energy requirements of the world's conurbations have led to areas of very high consumption density. Only rarely do these two concentrations coincide with the consequence that spatial networks may be very extended. Biomass-based systems, however, have quite different spatial characteristics with low consumption densities, dispersed supply and local scale networks.

Environmental The environmental component of the total energy system refers to the interactions between the stages of the respective energy chains and the biophysical environment. All stages unavoidably have an impact on the environment in proportion to the type and scale of the activities and the sensitivity and resilience of the biophysical systems involved (Fowler 1975; Holdren *et al.* 1980). Most stages of large-scale energy chains are visually obtrusive, resource consumptive (excluding hydro) and, to some extent, environmentally disruptive. The converse of this relationship is the influence of the environment upon the characteristics of the energy system. Many aspects of

9

the design and operation of the facilities which make up energy chains are a reflection of environmentally imposed conditions.

Social   The energy systems adopted by societies are a reflection of and have a penetrating influence upon, the values and structure of those societies. Thus, the influx of labour during the construction phases of energy projects has significant implications for the local population. The presence and operation of major facilities may affect the work experience, health and safety of both employees and the general population nearby (Budnitz and Holden 1976; Maxey 1980). More broadly, the general character of the prevailing energy systems has fundamental relations to the settlement pattern and transportation network (Nader and Beckeman 1978).

Variables

While the structure of energy systems may be *described* at any given time or place in terms of the preceding six components, a deeper understanding or *explanation* of the current structures and *prediction* of future structures requires a knowledge of the variables which influence them. These variables have been organized into six classes for discussion purposes (Fig. 1.4).

Technology   The existence and character of particular energy chains depends in a fundamental way upon the state of scientific knowledge and the development of technologies to put that knowledge to use. Thus, at all stages of all energy chains, machines, equipment and operating techniques depend upon the stage of techno-scientific development and its diffusion into general use (Lihach 1982). Some systems, such as those based on nuclear energy or synthetic fuels, are extremely complex in this respect and require the highest levels of skill and investment (the so called 'hard' energy systems). Others are simple and accessible to relatively untrained users (the so called 'soft' systems). It should be noted, however, that improvements in physical efficiencies, cleanliness and safety of the 'soft' systems depend to a considerable extent on the success of technological development of new materials and the design of innovative equipment (e.g. active solar systems). Furthermore, the transition from the present reliance upon conventional oil supplies to other sources is dependent to a very large extent upon new technologies (Barker 1981).

Benefits and costs   The energy systems which are adopted for widespread use are those which promise the greatest stream of benefits in relation to costs. The determination of economic efficiencies requires the consideration of such variables as cost–price relationships, elasticities, returns on investment and, more broadly impacts on economic structure, balance of trade and capital flow (Eden *et al*. 1982). Because of the difficulty of incorporating intangible items into calculations of benefit/cost ratios, those systems which demonstrate the greatest economic efficiencies are frequently selected (Copp and Levy 1982). Few would disagree that benefits should exceed costs but when these are measured only in economic terms, questions are increasingly raised about social and environmental aspects of the relationship (Winner 1982).

Regulation Decisions about what energy sources to develop, what technologies to adopt and what prices to charge have been made in most of the world by the energy industry itself in the context of market forces. However, the objectives and programmes of the large organizations concerned have not necessarily coincided with the interests and socio-economic goals of the societies they serve. With some notable exceptions (e.g. USSR), governments tended to take a 'laissez-faire' attitude toward energy developments until approximately the middle of this century when they became increasingly involved and, in many countries, government policies became the dominating influence on the energy scene (Ward 1974). In the 1980s, however, a reverse trend emerged with a move towards 'privatization' and the substitution of market forces for regulatory intervention.

Policies are effected by means of the regulation of and direct participation in energy matters. By regulation, government may set the operating conditions for some or all aspects of the industry and thus steer investment, determine rents and price structures, and set conditions for environmental protection. By direct participation, governments can place particular energy chains or parts of energy chains under their immediate control thus securing a 'window' on the industry as well as an operational presence. However, it must be remembered that policies of governments at one level are subject to influences from the larger political and economic systems of which they are a part. These contextual influences, plus tensions between regional interests within countries and changes of government may make policy conditions as changeful as those of the market.

Values and perceptions There is little doubt that technological, economic and political forces together have been dominant in determining the character of energy systems for much of the developed world. Decisions have been made on the basis of techno-economic criteria by very large, centralized organizations. Even with the increase in the role of governments, criteria which reflect such basic elements of the social fabric as equity and accountability and community participation have not played a notable role in decisions affecting the major characteristics of energy systems to date.

Recently, however, in many democratic societies organized social interest groups advocating wider concern for these elements are beginning to have a significant effect on energy developments. Frequently such groups seek to delay the need for further expansion of energy systems by advocating 'conservation' and to deflect future investment from the path of 'hard', centralized systems dependent for the most part upon non-renewable energy sources, to 'soft', decentralized and participatory systems based upon renewable resources (Craig *et al.* 1976; Gerlach 1982; Maxey 1980; Morrison and Lodwick 1981).

Resources and Residuals The relation between the biophysical environment and the characteristics of energy systems is complex and inter-active. First, consider the direct influence of the environment upon the energy system. Commonplace as it might seem, it is worth reminding ourselves that the character of the past and present environment largely determines the existence and quality of primary energy sources. Thus, geological and geophysical conditions determine whether fossil fuels are present, the hydrologic regime and landforms whether a hydro-electric potential exists and the climate–soil characteristics underlie the basic productivity of biomass. In other words, the environment primarily determines the basic energy resource opportunities available within a country or region. From an operational point of view all stages of energy chains may be affected by unfamiliar or extreme environmental conditions requiring the development of new technologies or procedures in order to avoid service delays or interruptions.

Environment also affects the potential demand for energy. The seasonal temperature regime determines the general level of the potential demand for heating and cooling built space while short term temperature extremes may lead to dramatic peaks of demand. Such abrupt temperature changes can place a heavy strain on energy distribution systems and require production systems to include expensive peak load and storage capacity.

Reversing the relationship and considering the effect of energy systems upon the environment, it has already been noted that every stage of the energy chain, regardless of the energy source involved, has some impact on the environment. The realization of this by relatively large segments of society in the developed world has led governments to establish increasingly stringent regulations which significantly increase the lead time required to introduce changes to existing energy systems and, in many cases, increase their cost.

Distance and density The effects of distance and density upon the structure of energy systems are also complex and interwoven. Generally speaking the global centres of energy consumption no longer have sufficient regional energy sources to meet their requirements. While it is true that some of the largest consuming areas grew up around coalfields, the general expansion of demand and the broadening of the types of energy required by these areas have led to local resources becoming at best inadequate and, more commonly, exhausted. As a consequence the frontiers of primary energy supply have been pushed ever more distant.

Decisions about the location of primary production facilities have to be made within the constraints of the distribution pattern of known resources. The more distant these sources are from the major markets the larger in size and/or higher in quality they must be if they are to be selected for production. Linear, physical measures of distance are not the most important in this assessment. Rather, it is distance measured in terms of costs and reliability of supply. Thus both the mode of transport and the routes followed play a greater role than sheer physical distance. For example, easy access to bulk water transport facilities may favour a physically more distant fossil fuel source over one, which although closer to the market, is dependent upon overland transport facilities. This is particularly the case if the access is to bulk ocean transport where not only are the energy/km costs lower but the producer has the advantage of being able to serve a greater range of delivery points. Fixed-route overland facilities do not provide such flexibility.

The high degree of resource-orientation which characterizes the location of primary production facilities is not found in the processing, transformation and storage phases. Rather these facilities will be located at that distance from the primary production site and market that yields optimum financial returns. For some facilities the optimum location may be market-oriented (e.g. many oil refineries) and, for others, resource-oriented (e.g. natural gas processing plants). However, the distance which is economically desirable may lead to a location which is unattractive (or not available) when assessed in terms of other criteria (Nader and Millerton 1979). Thus, for example, areas with high population density are likely to be the locus of loudly expressed concerns about safety and visual blight. Conversely, uninhabited areas distant from population centres may be prized as ecological preserves or areas of great natural beauty. As one writer recently observed the spatial dimensions of energy systems are those in which '. . . location decisions are frequent, often contentious and . . . the point at which differing views must be reconciled' (Wilbanks 1985, p. 219).

Population or, more precisely, energy consumer, density also plays a role in the spatial structure of energy systems. High densities of consumers require and are able to sustain large and numerous supply facilities built to take advantage of economies of scale. Low densities, on the other hand, may not afford sufficient demand to bring the supplier's revenue above a minimum cost-recovery threshold. The latter situation is particularly relevant in the provision of gas and electricity to rural areas where revenues per kilometre of distribution facility are low.

Energetics

The study of the structure and variables of energy systems was for a long time exclusively in the domain of the basic sciences as the fundamental properties of energy and matter were discovered and developed. Particularly in the context of thermodynamics, these studies focused upon the inputs, transformations and outputs of energy in physical and chemical systems under the general heading of energetics. By the 1970s the use of the term energetics had become extended well beyond the thermodynamics of particular equipment and processes to the flows of energy in biological and socio-economic systems of widely ranging scales and complexity.

Disciplinary roles

In this wider context each discipline tends to concentrate its attention upon those components and variables which most readily fit into its perceived subject matter and prevailing research methodology. Here the role of the major disciplinary groups in the general field of energetics is summarized leaving discussion of geographers' interests in the field to the next section.

Basic sciences Basic scientists in physics, chemistry and allied fields, provide the fundamental knowledge about energy and over the centuries they have discovered and codified the major principles of thermodynamics, electricity and nuclear energy. Future development of new and advanced energy systems depends initially upon their research. Workers in this group of disciplines use physical and chemical units, manipulate their data mathematically in the framework of scientific theory and are accustomed to a high level of precision and accuracy. In addition to their crucial role in the advancement of fundamental knowledge and their contribution to the penetrating analysis of energy systems, basic scientists also contribute to forecasting. They provide data about what is physically possible in the production and consumption of energy as well as providing the scientific assessment of the potential of new and advanced energy systems.

Applied sciences The technology used in energy chains is the particular domain of the applied sciences. Engineers mobilize the results of scientific research and design and develop the necessary equipment. Beyond this they conceive and carry out the construction phases of energy projects of all kinds and subsequently become responsible for plant operation. Thus they play a large part in determining the technical efficiency and safety with which operations are carried out as well as being responsible for the conduct of in-plant energy audits and analyses. Like the physical scientists, engineers work with physical units and systems and draw upon deductive process laws for analysis and prediction.

Environmental sciences This group includes the earth, biological, atmospheric and hydrological sciences. The major contribution of the *earth* sciences (geology and geophysics) to energetics is focused on the study of the stocks of fossil and nuclear fuels in the terrestrial environment. Other disciplines look to the earth sciences for expert judgements about the physical availability (the resource base) of these fuels and the geophysical conditions which determine their potential producibility. The *biological* sciences have made considerable advances in their knowledge of the circulation of energy within natural and managed ecosystems and now play a major role in the assessment of the interaction between the human use of energy and the biosphere. *Atmospheric* and *hydrological* sciences provide the primary information about the resource base of such renewable energy supplies as solar radiation, wind, wave and hydro as well as assessment of the impact of the energy chain upon atmospheric and hydrological systems. Collectively, the environmental, or 'field', sciences

14

have only limited opportunity to use the experimental, laboratory style of enquiry and have to deal with a wide range of variables and scales in their analyses. Nevertheless, they work with biophysical units in a mathematical and statistical context and seek to use process laws and principles in forecasting.

Social and policy sciences The interest of the social and policy sciences in energetics only emerged on a significant scale after 1950. Economic and financial analyses of all aspects of the energy chain now play a major role in determining the feasibility of energy developments and helping to choose between alternatives. Seeking insights into the inter-relation between economic and energy systems is, however, the *economist*'s main goal within the general field of energetics. By analysing energy elasticities and the relations between economic structure and energy demand, economic research lays the foundation for the major role that economists play in forecasting energy demand. Insights from the *behavioural* sciences have become increasingly applied to the energy scene as the importance of social values and perception has grown. Their interests include the impact of energy developments upon social conditions, consumer attitudes to energy issues, public beliefs about energy systems and the social acceptance of risk. The institutional and organizational components of energy systems are approached by *political economists* and *policy* and *management* scientists.

Collectively, the social and policy researchers work with complex, highly aggregated socio-economic systems, the character of which is partly the result of interaction with biophysical systems. Monetary, commodity and demographic units are commonly used by social scientists but because of the macro-scale of their analyses and the associated large numbers of uncontrolled variables, accurate prediction is elusive despite the sophistication of the mathematical and statistical modelling techniques applied.

Each of these disciplinary groups works most comfortably on particular foci within the spectrum of energetics and adopts its own preferred methodology, set of analytical techniques and units of measurement. Consequently, really holistic and balanced enquiry in energetics requires an inter-disciplinary approach. The difficulty of bringing the different paradigms together in an effective way is well illustrated by the emergence of an approach to energetics first labelled energy accounting and, later, energy analysis.

Energy analysis

As individual physical and applied scientists extended their interests beyond physical systems to broader socio-economic concerns, they sought to apply their well-tested thermodynamic paradigm. Some, dissatisfied with social science approaches to economic issues, promoted the adoption of a 'scientific' energy based value system rather than one based on monetary values. By the late 1960s, analysts with their intellectual roots in the physical and the biological sciences were devoting themselves to the study of the human use of energy.

Two problems quickly emerged. First, what energy concepts, methodologies and units were to be used in order to make studies comparable? Secondly, what did this form of analysis add to the conventional economic analysis based upon monetary units and the operation of market forces? In particular, could an energy based theory of value replace or, at least co-exist with, the prevailing money based theory?

The first of these difficulties was addressed in 1974 at a workshop convened by the International Federation of Institutes for Advanced Studies (IFIAS). The workshop recommended:

> that the unit of account be defined as the amount of energy source expressed in joules, which is sequestered by the process of making a good or a service, and that it be named Gross Energy Requirement . . . (Nilsson 1974, p. 222).

Participants in the workshop emphasized the importance of establishing consistent boundaries to the systems being investigated and identified the concept of Net Energy Requirement. The workshop also addressed the use that could be made of energy analysis and agreed that:

> . . . the price system did not always embody sufficient information to make decisions, or to make them in time. Energy represents a more sensitive signal of impending change . . . (Nilsson 1974, p. 274).

Subsequently other international and inter-disciplinary meetings took place and there was a flood of energy analysis literature, both professional and popular (*Energy Policy* 1975). Procedural conventions were refined (though important differences remained) and the clash between the basically thermo-dynamic approach of the bio-physical scientists and the econometric approach of the social scientists became more heated (Gilliland 1975; Huettner 1976). However, by the end of the 1970s the debate had subsided and each side had recognized the strengths and weaknesses of both approaches. Although individuals continue to hold extreme views, the consensus now seems to be that, as in most human affairs, decision making benefits from considering evidence from several value systems rather than depending on only one.

Geography and energetics

The primary interests of geographers are centred upon the study of the biophysical environment, the spatial organization of human activity in that environment, and the inter-relations between the two. Geographers were quick to appreciate that the human capacity to do work and, therefore, human control over energy, is a crucial element in a study focused on man and environment. Prior to 1950, despite this broad appreciation of the role of energy, geographical studies were mainly limited to descriptions of the spatial distribution of individual energy sources and primary energy production at various scales (Chapman, J. D. 1961).

In the next fifteen years, however, both the multi-component structure of energy systems and the importance of thermodynamic principles were recognized. Three relatively comprehensive books on the economic geography of energy were published (George 1950; Chardonnet 1962; Manners 1964) and the environmental science side of geography moved towards a dynamic, process-oriented approach which paid particular attention to energy flows (Hare 1953; Thornthwaite 1961; Linton 1965).

Thus, by 1965, the stage seemed set for two of geography's mainstreams to contribute to the emerging broader study of energetics. To a considerable extent this promise has been fulfilled by physical geographers (Chapman, J. 1976). Their studies of the biophysical environment have become solidly rooted in an approach which relies heavily upon an understanding of the flow of energy and materials as a determinant of the amount and timing of work in environmental systems (Carter *et al*. 1972).

Despite this relatively rapid adoption of modern approaches to energetics by the environmental science group in geography, recognition by human geographers lagged. Particularly in comparison with writers in other social science and humanistic disciplines, relatively little was contributed by geographers to the socio-economic aspects of energetics during the 1960s and 1970s. Some noteworthy books (e.g., Odell 1963, 1969 and 1970; Guyol 1971; Curran 1973; Wagstaff 1974; Chapman K. 1976; Cook 1971, 1976 and 1977; Dienes and Shabad 1979) and journal articles (e.g., Luten 1971; Osleeb and Sheshkin 1977) appeared but geographical research programmes in university departments, special interest groups or journals focusing on the geographical treatment of energetics in a socio-economic context were slow to emerge.

By the mid-1980s the scene had changed somewhat. Calzonetti and Solomon (1985) reviewed progress in what they call the 'geographical dimensions of energy'. After a review of geographical writing on each of the major energy sources and conservation, the book presented nineteen commissioned papers under four headings considered to represent the current foci of socio-economic geographers' interests in energetics:

1. resource development issues
2. power plant siting and land use
3. patterns of energy use
4. multi-regional and environmental issues

In the concluding chapter, Wilbanks addressed the role and mission of geography and geographers in the study of energy and their potential contribution to policy analysis and formation (Wilbanks 1985). In this chapter, Wilbanks suggested that the likelihood of geographers being able to solve future energy problems '. . . depends considerably on our ability to identify and understand the workings of energy in our societies and economies.' (p. 506). Later he referred to the spatial awareness of geographers and their familiarity with '. . . location, spatial structure and movement' (p. 507) and as a consequence, their ability to produce useful advice to policy makers on location decisions, movement systems, spatial structures and the relationships between people and their environment (p. 508).

It is intended that this book shall contribute to the development of understanding of the working of energy systems, to a knowledge of the spatial structure of energy systems, and an introduction to the components of public policy and management. Spatially and temporally the empirical information is presented at a relatively high level of generalization, dealing mainly with major international economic groups and individual countries over decades rather than intranational divisions or particular siting problems. The conceptual frameworks that are reviewed are also broad to match these general spatial and temporal scales as well as to provide a reasonably comprehensive basis from which to approach more immediate and localized energy issues. The functional and other structural components of both the consumption and supply side of commercial energy systems are presented before the spatial aspects are dealt with. The focus is upon large-scale commercial energy systems rather than traditional or emerging systems which deserve separate treatment in another volume. The book is intended not only for the geographic reader but also for those with other backgrounds who might find material and ideas which will add to their understanding of the energy scene.

CHAPTER 2

Energy consumption: components, variables and profiles

In private life and at the work place, each of us is an energy consumer and the aggregation of millions of individual decisions determines the energy consumption of a society. The size and structure of energy consumption systems is thus very complex, subject to many variables and difficult to change quickly. This chapter considers the basic components of energy consumption and the variables which affect them. It also describes energy use in the major sectors.

Components

Energy consumption may be measured in terms of the type of energy consumed using *commodity* units such as the tonne, barrel or cubic metre or *energy* units such as the British Thermal Unit (BTU), joule or watt (Appendix A) and its structure may be analysed in terms of consumption categories (Fig. 2.1). Data on energy consumption are not as comprehensive or available as supply data, chiefly because of the difficulty of collecting quantitative information from the

Type of energy consumed	Consumption categories		
	Sector	Product system	Function
Solid			
Liquid			
Gas			
Electricity			
Other total			

Figure 2.1 Dimensions of energy consumption

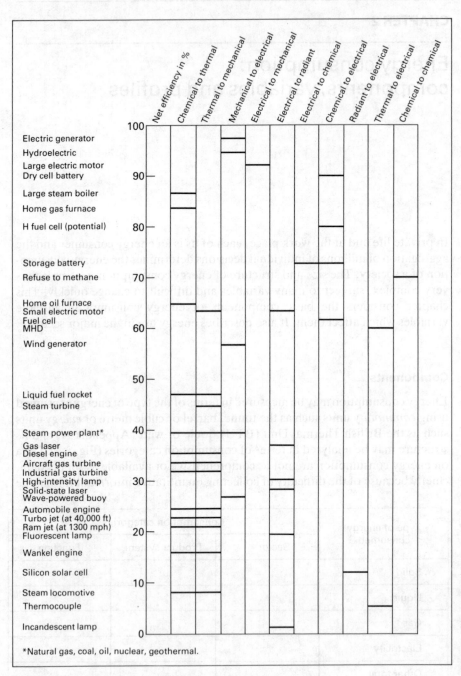

Figure 2.2 The net efficiency of selected energy convertors
Source: Fowler 1984

multiplicity of energy users. The data sets that do exist originate from the supply side except in the relatively few instances of special consumer surveys. The total energy consumed by any economic system is statistically defined as being the disappearance of total energy supplied less exports, bunkers and storage. Initially the data are in commodity units which are then converted into a common energy unit and summed. When comparing such summed figures care must be taken to ensure a compatibility of conversion factors and conventions and that double counting is avoided.

In addition to these gross measures of energy consumption there are three general classes of derived measures: energy efficiency, intensity and density. *Energy efficiency* refers to the ratio of desired work (or energy) output in relation to energy input ($\times 100$) and is used mainly in a thermodynamic sense when one type of energy is changed to another in various convertors (Fig. 2.2).

The amount of energy used per unit of output is measured by *energy intensity* or *energy cost* (E/W or the reciprocal of efficiency) where output may be expressed in commodity, fiscal or, in the case of transportation, weight $\times$ distance units. The consumption of energy per unit area (terrestrial surface or built space) yields the third derived measure, *energy consumption density*.

Consumption categories

Knowledge of the total amount of energy consumed provides no information about the work which it performs. To document the work carried out, energy must be traced to a consuming sector, a product system or the function performed.

Sectoral Most consumption data sets identify four or five major consuming sectors as illustrated in Table 2.1. For more detailed analysis it is useful to

Table 2.1 Commercial energy consumption by economic sector: World and regions, 1972

	North America	Western Europe	Japan Aust.	USSR and E. Europe	Developing countries*	World
			Per cent			
Transport	23	14	18	9	24	18
Industry and agriculture	22	26	32	45	38	31
Residential and commercial	21	27	17	15	17	20
Non-energy uses	6	6	9	1	3	4
Energy conversion and losses	28	27	24	30	18	27
Total	100	100	100	100	100	100

* Includes China and centrally planned Asia
Source: Eden *et al*. 1982

subdivide the major sectors further (as in the profile section of this chapter) and to distinguish between the intermediate and final use of energy. The former refers to energy used as an input in the production of goods and services for sale such as in agriculture, manufacturing or public transportation. Final use refers to the energy consumed to produce goods and services which do not enter the market such as in subsistence agriculture, household operations or personal transport.

The transport sector includes all for-hire transport services as well as personal transportation while the industrial sector refers to all categories of factory manufacturing except the energy conversion industry which is usually identified separately. The agricultural sector, listed in Table 2.1 with industry, is sometimes shown separately or grouped with residential-commercial. The residential or domestic sector refers to final energy use in the home and the commercial sector to intermediate use in buildings used to provide such functions as wholesaling, retailing and education. The non-energy use category refers mainly to the use of energy commodities as feedstock for the petro-chemical industry and as lubricants and waxes.

Product system Energy consumption may also be analysed in terms of product systems which involves tracing the energy used in all stages of the production, transportation and final consumption of a particular product (Taylor 1982). The product system form of recording energy consumption is complex and requires special scientific and engineering procedures. In such studies it is necessary to distinguish between the use of embodied (sequestered) energy and operating energy. Embodied energy refers to the energy required to manufacture the capital equipment used to provide a product, while operating energy is that used in the actual operation of the equipment. In this

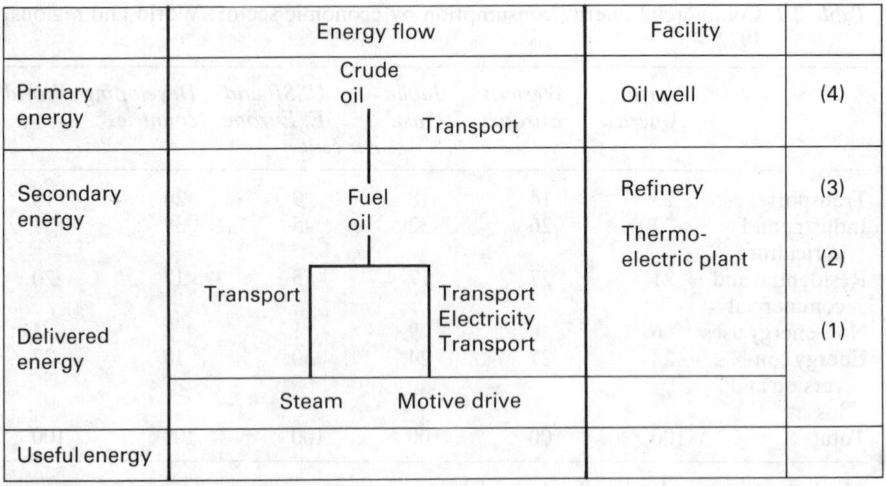

	Energy flow		Facility	
Primary energy	Crude oil	Transport	Oil well	(4)
Secondary energy	Fuel oil		Refinery	(3)
			Thermo-electric plant	(2)
Delivered energy	Transport	Transport Electricity Transport		(1)
	Steam	Motive drive		
Useful energy				

Figure 2.3 Bounding the product system

Table 2.2 End use of energy: USA

	Residential	Commercial	Industrial
	(Percentage of total energy used in each sector)		
Space cond.	61.2	60.4	—
Water heating	14.9	7.5	—
Cooking	5.5	1.6	—
Refrigeration	6.0	7.6	—
Process steam	—	—	40.6
Direct heat	—	—	27.8
Appliances	6.6	—	22.0*
Asphalt and oil	—	11.2	—
Feedstock	—	—	8.8
Other	5.8	11.7	0.8
Total	100.0	100.0	100.0

* Motors and process
Source: Fowler 1975

form of analysis it is important to identify clearly the extent of the system being considered. It may be seen from Figure 2.3 that if only energy delivered to a plant gate is recorded (1), no account is taken of the energy used to generate the electricity (2) or in the case of the fuel oil, to refine (3) or produce (4) the crude oil (Bush and Chadwick 1979). Furthermore the energy cost of transportation between the stages of the energy chain, the energy embodied in materials processed and the energy involved in the delivery system for the product, are not included. When only the delivered energy is considered it is usually referred to as Net Energy used and, when one or more of steps 2–4 in the energy chain are included, as Gross Energy used.

Function Energy consumption by function usually refers to the amount of energy used for such purposes as transportation, space heating, lighting and cooking. Data organized in this form are not generally available from annual statistics though most developed economies conduct periodic surveys and publish the results in occasional reports (Table 2.2).

For many uses heat is the final form in which energy is consumed and it is useful to know in what heat ranges energy is required. Table 2.3 illustrates, for Canada, that more than three-quarters of the heat consumption is for low temperature uses. To the extent that the heat energy in the energy supply exceeds that required by the function to be performed, there will be residual energy which either has to be disposed of or is available for re-use.

Type

The analysis of energy consumption is still only partially complete when the total amount is known by consumption category. For complete documentation

Table 2.3 Temperature distribution of energy consumed as heat: Canada

Temperature	10^{15} joules					
	Energy supply industry	Domestic sector	Commercial sector	Industrial sector	Total	%
High temperature (>260 °C)	47	0	0	194	241	8.9
Intermediate temperature (140–260 °C)	70	0	0	250	320	11.9
Low temperature (100–140 °C)	126	0	0	457	583	21.7
Space heating (<100 °C)	14	905	580	50	1549	57.5
Total	257	905	580	951	2693	100.0

Source: Puttagunta 1975

it is necessary to know the type of energy consumed (Guyol 1971). This involves recording the general form (e.g. solid, liquid) and specific type (e.g. coking coal, heavy fuel oil) of energy distinguishing between primary, such as natural gas, or secondary, such as thermally generated electricity.

Variables

The consumption of energy derives from the goods and services which its use provides. Thus the amount and composition of energy consumed is a function of the general demand for these goods and services which, in turn, is subject to the influence of many variables, some major and fundamental and others of importance over a restricted range of space and time. The four major variables and their associated sub-variables are shown in Table 2.4.

Table 2.4 Variables influencing the consumption of energy

First order	Second order	Third order
1. Level of economic activity	Structure of economy	Technology and equipment stock
2. General price level of energy	Difference in price between energy sources	Regional variations in price
3. Population – numbers	Income settlement patterns	Cultural characteristics and lifestyles
4. Environment – temperature regime	Environmental regulations	Spatial extent

Economic activity

The general level of economic activity is the major variable determining the total amount of inanimate energy consumed (Eden *et al*. 1982). During periods of rising production, trade and consumption of goods and services the overall demand for energy also rises. The relation between energy consumption and GNP (or GDP) has been extensively investigated and although robust descriptions of the relationship are elusive, the expected, generally positive relationship has been clearly documented.

Within this major variable, the structure of the economy and the technology and equipment stock are important sub-variables. In the first instance, economies with large manufacturing sectors (particularly primary processing) are likely to use more energy per unit of GNP than those in which the tertiary or quaternary sectors are dominant. Secondly, the thermodynamic efficiency with which work is carried out in an economy (and thus the amount of energy required for a specific task) is determined by the technology used, the character of the equipment stock and the manner in which it is operated.

Price

A second major factor determining the consumption of energy is its real price in comparison with other factors of production. If the real price of inanimate energy declines relative to labour or capital, there will be a tendency to substitute it (with a time-lag) for the other inputs in the production of intermediate goods and services. Similarly, in the case of final goods and services the price of energy in relation to disposable income will play an important role in determining the mix of human labour and inanimate energy that is used. If we can afford it, most of us are inclined to minimize our own labour input by substituting some appliance which requires inanimate energy. Detailed analysis of the operation of these variables confirms the expected general tendencies but precise relationships of the sort which will permit confident prediction have been elusive (UK, DOE 1977a).

Population

It is an obvious, but often overlooked, fact that the energy consumed by a society is a function of the number of people times their wants and their ability to satisfy them. For each net addition to a population there is an added increment to the energy requirements of that population. How large that increment is depends upon income, settlement pattern and lifestyle.

Environment

The biophysical environment of an area influences the demand for energy in three rather different ways. First, the temperature regime is a major determinant of the amount of energy required for space conditioning although the actual amount used will be a function of such disparate variables as the level of insulation of houses and the cultural and lifestyle preferences of the population. Second, all links in the energy chain have some environmental consequences. To mitigate or control these consequences many societies have introduced regulations which may not only change the cost structure of energy systems but may directly determine primary energy production, input and processes. Less obvious, but still a consequence of the environment of an area, is the relation between sheer physical size (and thus the distances over which goods and personnel have to be transported) and the amount of energy required for transportation. Generally, it is to be expected that the amount of energy used for transportation in large countries (e.g. USA, Australia) would be greater than in small ones (e.g. UK, Switzerland).

Decision making

These four factors together constitute the major factors influencing the demand for energy at any given spatial scale and during any particular time period. They become operational as the result of the judgements and decisions of individual and group decision makers. Such decisions may be taken with the deliberate intention of changing the energy input or they may be taken in order to achieve some other objective but have an energy outcome, perceived or otherwise. Which of these two classes of decisions prevails depends to a large extent upon the energy awareness of the decision makers and their susceptibility to energy supply and costs.

Decision makers fall into three broad groups – individuals, corporations and government agencies. *Individuals* make energy related decisions as householders, as managers or operators of some enterprise, or as designers and builders. In these roles they make operational and investment decisions which singly may have little effect on energy demand but collectively have a major influence.

Corporate decision makers influence energy demand in proportion to the amount of energy used in their enterprise or to the energy-using characteristics of their products. Firms with large energy requirements (e.g. many primary material processing industries or long distance transportation operations) will be sensitive to the amount of energy they use, and, over time, adjust the mix of energy, capital and labour inputs in response to their relative cost. Firms engaged in the manufacture of energy convertors influence demand by the energy-using characteristics they build into their product (e.g. automobile manufacturers) while energy suppliers, concerned to maximize the sale of their product, develop marketing strategies which play a significant role in determining the amount of energy consumed.

The role of *government* decision makers in influencing energy demand has become increasingly important in recent years. As legislators or employees of government agencies of various kinds, they play the role of regulators. Thus their influence can be widespread since they may directly determine the price of energy, product specifications, the conditions influencing capital investment and the availability and conditions of use of different forms of energy. When energy-producing firms become publicly owned, government decision makers may also influence the supply of energy available for consumption. Finally, as operators of large building complexes and transportation fleets, governments with their own energy-using patterns can play a significant part in determining the total energy demands of an economy.

Profiles

As an introduction to the spatial distribution of energy consumption in the following chapter, this section examines some of the characteristics of energy consumption in a selection of the major consuming sectors. In the first two sectors dealt with, domestic and commercial, much of the energy is required to condition the buildings in which a wide variety of personal, social and business services is carried on. In industry, the third sector to be covered, energy is consumed mainly for material processing while in the fourth sector, transportation, it is used by a wide variety of engines to provide motive power.

Domestic

Consumption of energy in the domestic sector refers to that consumed at the household level, both within the residence and beyond it (Amman and Wilson 1980). The major energy-using functions in the household are food processing (cooking, preserving), space conditioning (heating, cooling and lighting), cleaning and maintenance (personal, built space, grounds), recreation and communication. Some elements of these items are common to every household and the variations that occur are the result of different cultural characteristics and lifestyle, income, available energy supplies, and, in the case of space conditioning, the prevailing temperature regime.

The second class of activity occurs beyond the house and includes the acquisition of food, water, fuel and the provision of personal transportation. Acquisition of these basic inputs is a major activity in the rural subsistence economies of the world and thus for more than half the world's population. The eternal task of providing the household with biomass or waste for fuel itself consumes a great deal of human energy and time. Similarly the provision of food requires considerable human effort which, in turn, is energized by the output of that effort – food.

Such activities do not significantly enter into the lives of the world's urban population nor of the rural and 'rurban' population of developed economies. Rather, water and energy are supplied by formal distribution facilities and food is purchased. The journey to shop becomes, with the journey to work and to recreation, one of the major generators of the demand for personal transportation. Despite the large amount of energy used by individuals for personal transportation, national agencies usually record this use under the transportation sector.

The energy used by a household is not only a function of the work to be done but also of the stock of tools, appliances and structures available to the user. The size, type and material composition of the house itself is a fundamental factor in determining the space conditioning requirements in any given ambient atmospheric conditions. The utensils and appliances used for processing food and personal hygiene vary from the open fire, basic stove and wash basin to the full range of appliances found in a modern detached house in a high income district of a metropolitan centre. The use of energy for domestic purposes in different societies is highlighted in Table 2.5 which, although based upon detailed data from several sources, presents composite profiles to illustrate the variation between three groups.

Rural developing The rural-developing profile represents a subsistence agricultural economy. The majority of energy is consumed beyond the home, producing food and providing water and fuel. The main energy source is metabolic which in turn is provided by the food and feed available – a vicious circle (Revelle 1976). If the productivity of agriculture is to increase it is likely that artificial fertilizer and perhaps, irrigation, will have to be used and both, directly or indirectly, require more energy derived from hydrocarbons or primary electricity. In-home energy use is minimal, almost entirely concentrated on the preparation of food and maintenance of clothing (e.g. ironing) and supplied from unprocessed biomass and animal wastes. In many countries the pressure on these sources has grown to the point that harvesting of firewood far exceeds the mean annual increment of growth in accessible areas with the result that the regenerative capability of the forested areas is being jeopardized on a large scale. Furthermore, the human effort that has to be exerted to find and transport supplies is increasing considerably (Wood and Baldwin 1985).

In this setting, energy-using domestic appliances are frequently limited to a stove (often very inefficient as a convertor) and a lamp burning vegetable oil or kerosene. Space heating is not required in most locations and cooling is usually limited to natural ventilation. Above all, poverty precludes any capital outlay for additional or replacement equipment even if operating energy costs could be met.

Urban developing The household consumption of energy in developing economies changes in a number of significant ways when an urban setting is considered. Most notable is the change in type of energy used from predominantly metabolic and raw biomass to processed fuels (first charcoal and then liquids) and electricity. The major reasons for this are (1) the substitution of piped water supply and commercial food and fuel sources for the subsistence activities required for these essential commodities in a rural setting and (2) increased income. For the low income family this shift is accompanied by a major relative increase in the proportion of total energy consumed in-home.

Table 2.5 Composite profiles of domestic energy consumption

	Rural developing	Urban developing		Urban developed
Annual income $	250	1–4,000	5–10,000	30–40,000
Total energy 10^6 BTU	75	25	110	300
Percentage of total used:				
in-home	25	85	30	60
beyond home	75	15	70	40
Percentage of in-home used for:				
space condit.	0	10	20	65
other	100	90	80	35
Percentage of total energy by type:				
metabolic				
human	20	—	—	—
animal	50	—	—	—
Biomass				
wood	7	—	—	—
crop waste	7	—	—	—
animal waste	14	—	—	—
charcoal	—	65	5	—
Fossil fuels				
solid	—	—	—	—
liquid	2	30	80	⎫
gas	—	—	—	⎬ 90
Electricity	—	5	15	10

Source: Makhijani and Poole 1975; McDougall *et al.* 1979; McGranahan and Taylor 1977; McGranahan, Chubb and Nathans 1979

Travel beyond home is limited to public transportation, walking or bicycling (the metabolic energy used in the latter is not included in Table 2.5). For the higher income family, however, beyond-home consumption re-emerges because of the ability to purchase one or more powered vehicles which results in the use of a considerable amount of energy for personal transportation. This and the use of energy for space conditioning as well as the operation of a growing number of basic and discretionary appliances accounts for the rapid growth of total energy consumed by the higher income group.

In view of the rapid urbanization of the population in developing countries these representative data for urban households have significant implications for both the amount and types of energy that will be required in the future. In particular they document how the domestic sector requirements for liquid fuels and electricity rise as urbanization increases and income grows.

Urban developed The mid-latitude, relatively high income profile indicates a large total energy consumption, a slightly greater proportion of energy used in-house than beyond, a major requirement for space-conditioning and a complete reliance on hydrocarbons and electricity. The large total in this profile results primarily from the energy used for private transportation (assuming one or more cars in the standard household) and for overcoming winter cold and, perhaps, summer heat. With a wide range of convenience and automated (e.g. hot water heating) appliances the in-home use is twice that of the low latitude, urban middle income household and three times that of the low income household. This difference is all the more remarkable when consideration is given to the amount of in-home functions that have been transferred to the commercial sector in developed urban economies (e.g. precooked foods, restaurants, laundries, cleaners).

Commercial

As a consumer of energy the commercial sector is defined in several different ways thus making it difficult to compare data from one source to another. In most cases the commercial sector is a residual category including all activities not included in other, separately identified sectors. Precisely what activities are included consequently varies from one data set to another and may include elements of both primary and secondary economic activity (e.g. agriculture and construction) as well as tertiary and quaternary. Furthermore, when deciding whether to record consumption in the domestic or commercial sector, the rate charged by suppliers is often used as the distinguishing criterion. Thus, if the consumption of energy by any customer is sufficiently large to qualify for a bulk rate (i.e. 'commercial' rate) that consumption will be assigned to the commercial sector though very large consumers qualify for industrial rates and are thus classified into the industrial sector. However, despite the variety of energy users that may be included in the commercial sector, it consists of a common core of basic services provided by tertiary and quaternary economic activity (Table 2.6).

Table 2.6 Major functions and facilities included in the commercial sector

	Function	Facility
Building	Administration	Offices
	Trade	
	wholesale	Warehouse
	retail	Store
	Public service	
	health	Hospital
	education	School, College
	Recreation	Theatre, Pool
	Accommodation	Hotel
	Personal service	
	food and drink	Restaurant, Bar
	cleaning and	Cleaners,
	maintenance	Auto repair shop
Non-building	Municipal service	Street lighting, waste collection, vehicles

Most commercial energy consumption data originate with the energy suppliers who do not readily have the information to disaggregate consumption into its constituent elements other than that provided by rate structures. Such disaggregation can only be accomplished by occasional, special studies which may be conducted at various scales (e.g. regional, urban, individual building) and be based upon specific user surveys, generalized extrapolation or simulation.

These special studies face major practical and conceptual problems. Among the former are those associated with the acquisition of actual consumption data at a useful level of disaggregation. In most cases it is necessary to have both special analyses of suppliers' delivery records and careful listing by users of energy consumed, equipment and appliance stock and operating procedures. Since the purpose of most studies of this kind is not only to describe energy consumption in the commercial sector but also to identify the operating variables, it is also necessary to gather data about building characteristics. This requirement often poses practical problems which may be partly solved by using municipal records but often requires building-by-building surveys.

The end use of energy consumed in the commercial sector is illustrated for Auckland, New Zealand in Figure 2.4. Space conditioning (including lighting) is the dominant use for all commercial functions followed by 'task-specific' uses which refers to the energy used by equipment to process information and materials (e.g. computing, reproduction, diagnosis, cooking). The transportation category refers to the energy used to provide vertical movement of personnel and material in multifloored buildings.

A graphic summary of the analysis of commercial use of energy in terms of four derived measures is shown in Figure 2.5 again for Auckland. The four most energy intensive commercial facilities are those in which health, office, accommodation and retail services are provided. On the basis of plant density, education replaces accommodation in the top four but drops to the bottom in utilization rate because of week-end and seasonal closure. The provision of health care in hospitals and, to a lesser extent hotel services, requires a high degree of internal comfort on a continuous basis and access to a variety of task-specific equipment. Studies in the USA of commercial sector energy use (excluding publicly controlled facilities, and using a more finely disaggregated facility classification) find recreational, retail, accommodation and office services to be the most energy intensive facilities (Bernstein 1975; US, OTA 1982).

In sum, the wide variation of energy use within the commercial sector results from two major groups of variables, user requirements and building characteristics. The first determines the levels of space conditioning and types of task-specific equipment required as well as the manner and frequency of their operation. Building characteristics include size, shape, construction materials and the character of the space conditioning equipment. Age of building has also been found to be a significant general factor because many energy-significant variables are embodied in the design characteristics of the building which in turn are often time-dependent.

31

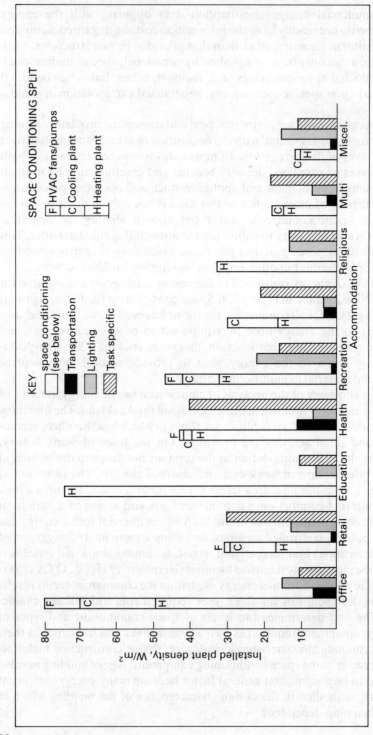

Figure 2.4 End use of energy in the commercial sector: Auckland, New Zealand
Source: Beca *et al.* 1979

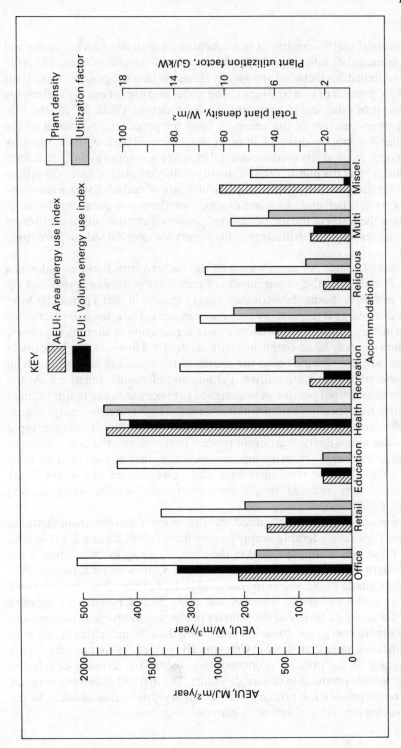

Figure 2.5 Energy density in the commercial sector: Auckland, New Zealand
Source: Beca *et al.* 1979

Industrial

The industrial sector consists of manufacturing industries which process and assemble materials into saleable products and non-saleable wastes. The wide variety of industries included are usually classified into subgroups on the basis of product, process or material input. The major end uses of energy and energy products within the sector generally are illustrated in Table 2.7 for the US. Almost three-quarters of the energy is used for processing and most of the remainder for motive drive of in-plant machinery. The feedstock category refers to energy products used as material inputs (e.g. naptha in the petrochemical industry) and the minor 'other' category mainly consists of space conditioning and service water heating. The particular mix of end-uses varies from one industry to another and, to some extent, even between plants in the same industry as the result of the use of different processes and the output of different product mixes. Figure 2.6 illustrates this variety for selected industry groups in the UK.

The mix of sources of energy used by this sector varies between industries (Fig. 2.7) reflecting the requirements of particular processes, price and reliability of supply. Some industries are energy specific in that a particular form or source of energy is required by the processes used. Thus, the manufacture of pig-iron in blast furnaces requires coke and the reduction of alumina to primary aluminium can only be accomplished with electricity. However, the majority of industries are less specific about the source of their process heat although all require electricity for motive drive, lighting and electronic functions. At the outset most industrial plants can be designed to make use of any major primary heat source but, once the equipment is installed, a shift from one energy source to another usually involves significant capital outlay which inhibits rapid change. The major energy carrier for process heat is steam, the heat content of which, once discharged from the primary operation, may be used in a cascading fashion (Fig. 2.8) for combined heat and power cycles and other lower temperature uses such as space conditioning and service water heating (Canada, EMR 1981).

The amount of energy required by the many different manufacturing industries varies considerably as may be seen from Tables 2.8 and 2.9. The data in both these tables are expressed in the form of energy intensity, that is the ratio of energy input to product output. Table 2.8, drawing on Canadian data (Statistics Canada 1978), shows the major industrial groups classified into three broad categories of energy intensiveness using monetary units to measure output. Category I consists of the primary processing industries which require five times more energy per dollar of value added than the industries in Category II and ten times more than those in Category III. Table 2.9 illustrates the energy intensiveness of the primary production and secondary recovery of selected energy intensive products in commodity units. The marked difference between energy requirements for primary and secondary production testifies to the energy-saving potential of increased material re-cycling.

Table 2.7 End use of energy in the US
           manufacturing sector

Process		71.2%
Steam	40.6	
Direct	27.8	
Electrolytic	2.8	
Motive		19.2
Feedstock		8.8
Other		0.8
		100.0

Source: Fowler 1975

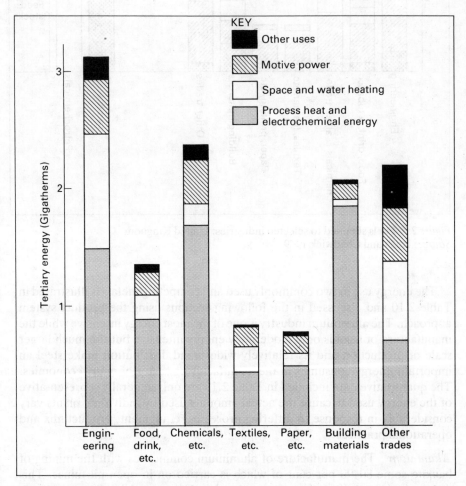

Figure 2.6 End use of energy in industry: United Kingdom
Source: Bush and Chadwick 1979

35

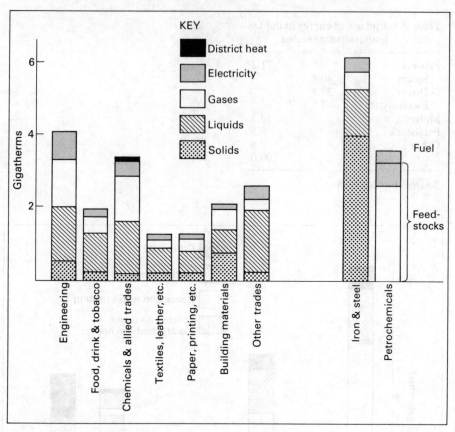

Figure 2.7 Fuels supplied to selected industries: United Kingdom
Source: Bush and Chadwick 1979

The energy use in two commonly used and competing metals is illustrated in Table 2.10 and discussed in the following sections using the product system approach. The aluminium industry is one of the most energy intensive while the manufacture of steel is only moderately energy intensive but the much larger scale of production and its relatively widespread distribution make steel an important energy consumer in the manufacturing sector of many economies. The quantitative data included in Table 2.10 are only generally representative of the energy used because the actual amounts used by individual plants vary considerably in response to differing processes, equipment, product mix and operational practices.

Aluminium   The manufacture of aluminium commences with the mining of bauxite some 80–90 per cent of which is carried on in open-pit mines. This activity involves large volumes of material and the use of draglines, scrapers, trucks and conveyor belts all of which require mechanical energy, the primary

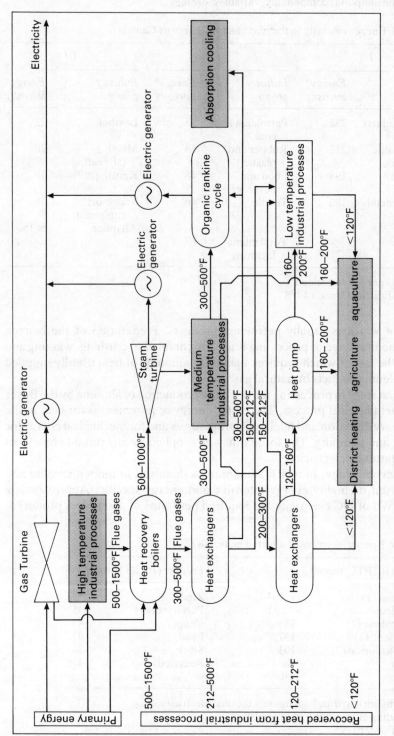

Figure 2.8 Illustration of an ideal energy cascading system
Source: Canada, EMR 1981

Table 2.8 Energy intensity in the manufacturing sector: Canada

I		II		III	
Industry group	*Energy* intensity*	*Industry group*	*Energy* intensity*	*Industry group*	*Energy* intensity*
Paper products	232	Petroleum and coal	53	Leather	22
Non-metallic minerals	217	Rubber and plastic	53	Metal fabricating	20
Chemical	198	Food and beverage	48	Knitting mills	19
Primary metals	186	Textile	46	Transport equipment	16
				All other	5–15
		Total manufacturing	67		

* 10^3 BTU per $ value added
Source: Statistics Canada 1978

source of which is usually petroleum products. Preparation of the bauxite before the first stage of processing is usually confined to crushing, washing and drying, the last of which requires significant amounts of heat usually supplied by petroleum products or natural gas.

The first stage in processing the ore is the production of alumina by the Bayer hydro-metallurgical process. The major energy requirement is for heat in the preliminary digestion and final calcination stages and for mechanical energy for grinding and pumping. The former may be supplied by any suitable fossil fuel and the latter by electricity.

The second stage, in which aluminium is dissolved in molten cryolite and electrolysed, is highly energy intensive and energy specific. Approximately 14,000 kWh of DC electricity are required per tonne of aluminium produced.

Table 2.9 Energy intensity of selected materials

Product (10^6 BTU/tonne)			*Product* (10^6 BTU/tonne)	
Aluminium (1)	75	167	Zinc	52
Recycled	5	11	Polystyrene	45
Polypropylene (2)	145		Copper	40
Polyethylene (2)	137		Lead	31
Polyvinylchloride (2)	103		Steel	26
Cut glass	80		Recycled	4–6
Porcelain	80			

(1) Low figure – hydro; high figure – thermal electricity
(2) Including energy content of feedstock.
Source: Fowler 1975

Table 2.10 Energy use in the primary production of aluminium and steel

Aluminium

Function	Form	Source	Amount*	Total*
Primary production				
Bauxite mining	Mechanical	Petroleum products, Electricity		
Preparation				
Ore drying	Heat	Petroleum products, Natural gas, Coal	30	30
Material processing				
Stage I — Alumina	Heat	Fossil fuel or Electricity	30	60
Stage II — Aluminium ingot	Electricity	Hydro or Thermal	230	290
Stage III — Shaping	Heat and mechanical	Fossil fuel and Electricity	25	315

Steel

Function	Form	Source	Amount*	Total*
Iron ore mining	Mechanical	Petroleum products, Electricity	0.2	0.2
Beneficiation Agglomeration	Mechanical, Heat	Fossil fuel, Electricity	4.0	4.2
Pig iron	Heat	Coke	26.2	30.4
Steel ingot	Heat, Electricity	Fossil fuel, Electricity	1.4	31.8
Shaping	Heat, Mechanical	Fossil fuel, Electricity	3.2	35.0

* GJ/Tonne of finished metal
Sources: Hannon and Broderick 1982; Kakela 1978; Taylor 1982; UK Department of Industry 1979

The gross energy requirement (GER) for this stage depends on whether the electricity is generated from hydro or thermal sources. The 230 gigajoules per tonne (GJ/t) shown in Table 2.10 is representative of the GER for thermally generated electricity but does not include a further 12–18 GJ/t required for the processing and controlled disposal of fluoride fumes and other wastes.

The liquid output of the aluminium reduction phase is generally cast into ingots which, when cooled, are available for shipment to shaping plants. The shaping stage usually requires heat for re-heating prior to the rolling and extrusion processes. These processes also require significant amounts of electricity for driving the rollers and rams.

Steel   The principal material inputs into the manufacture of steel are iron ore and coking coal. After mining, the majority of iron ore is prepared by sintering or, increasingly, pelletizing before being shipped for smelter use. Sintering is a roasting process which requires substantial quantities of heat whereas pelletizing is an agglomerating and hardening process that requires both heat and electricity.

Coal, in the form of coke, acts not only as the major energy source for the first pig-iron stage but also plays crucial physical and chemical roles in the overall operation of the blast furnace. Both the coke ovens (which convert coal to coke) and the blast furnaces produce considerable amounts of by-product gases which are used for a variety of in-plant purposes. The general trend of coke consumption per tonne of iron produced is declining as the result of the increased preparation of the ore and improved operating practices. Nevertheless, the coke requirement of blast furnaces is another example of an industrial process that is energy-specific. Direct reduction processes, increasingly used for small scale production of pig-iron, while individually energy specific, do not depend upon coking coal.

Conversion of the pig-iron into steel is largely carried on today in Basic Oxygen Furnaces (BOF) and electric furnaces. The former are charged with molten pig-iron and some scrap, while the latter are charged with scrap, usually cold. The BOF furnace requires considerable amounts of heat supplied by fuel oils and natural gas and 'captive' energy from the coke oven and blast furnace operations. Electric furnaces, on the other hand, are energized by electricity and thus are specific in their requirements of secondary energy. The proportion of steel-making capacity in these two major types of equipment determines the ability of the industry to use scrap and thus benefit from the attendant energy savings (Hannon and Broderick 1982).

Liquid steel from the furnaces may be cast in moulds and cooled into ingots only to be re-heated, at a relatively high energy cost, before being shaped into semi-finished products. Energy for the re-heat stage is reduced by the use of continuous casting in which the molten steel is immediately formed into semi-finished shapes such as billets and slabs. These shapes are in turn further shaped into finished steel products such as coils, bars and rods. Even with continuous casting, however, the final shaping usually requires some energy for re-heating as well as considerable motive power to drive the very large high-speed rolling and forming machinery.

Transportation

The transportation sector embodies all those facilities involved in moving personnel and goods (including energy) from place to place. The distinction between the passenger and goods elements of the sector tends to become increasingly marked as more specialized facilities are developed. Thus the automobile has become the most favoured vehicle for personal transport, the bus and commuter train for public transit, while trucks, unit-trains and specialized marine carriers have come to dominate the movement of goods.

The transportation function is provided by a variety of modes (Table 2.11) and requires an elaborate and capital-intensive infrastructure of routeways, traffic control and terminals. In addition to describing transportation systems in terms of modal mix it is customary also to describe them in terms of length of haul and organizational structure. Thus, distinctions are made between short and long distance hauls, feeder/distributor and trunk facilities which lead to such spatial categories as intra-urban, inter-city and international systems. Organizationally, there is a sharp division between the individual private ownership of personal vehicles (e.g. cars, vans, cycles) and public and corporate ownership of fleets of vehicles (e.g. public transit, airlines, shipping companies). The role of designers and operators in making decisions about vehicle characteristics and operating practices is becoming increasingly regulated by regional, national and international agencies.

It is estimated that approximately 20 per cent of the commercial energy consumed in the world is used by the transportation sector. Land modes are thought to account for about 75 per cent (road 60 per cent, rail 15 per cent), water 15 per cent and air 10 per cent and approximately two-thirds of the energy consumed is used to transport passengers (Eden *et al.* 1982). Unlike the other consuming sectors referred to in this chapter, the whole sector is very energy-specific, relying on petroleum products for more than 90 per cent of its needs and emphasizing the importance of the transportation sector in determining the demand for petroleum products.

Table 2.11 Modes of transportation

Medium	Land	Water	Air
Mode	Road – auto, motorcycle	Boat	Aircraft
	– truck, bus	Barge	LTOL
	– carts, wagons		STOL
	– bicycle		VTOL
	Rail – steam, diesel, electric		Dirigible
	Pipeline – variable diameter and pressure $\longrightarrow$		
	Transmission line – AC/DC, variable voltage		

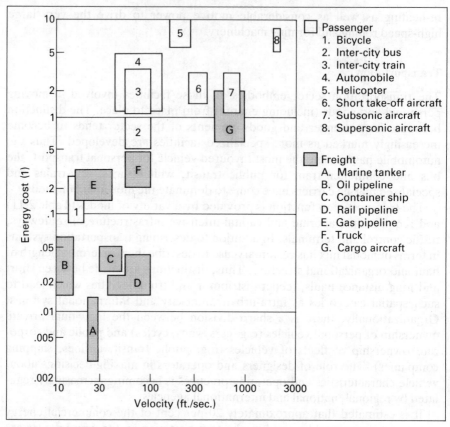

Figure 2.9 Energy cost of selected modes of transportation
Source: Cockshutt 1973

Measures of the amount and type of energy used by transportation are regularly available for all developed economies and a comparison of the relative importance of the sector in the total consumption picture reveals some wide variations between countries (Chap. 3). In order to understand these variations it is necessary to go beyond the basic measures of energy consumed in the sector and consider the amount of energy required to accomplish a given amount of transport work by various modes. Figure 2.9 shows the energy cost of various modes of transportation in relation to the general velocity range in which they operate, both expressed on a logarithmic scale. This figure illustrates three major characteristics of energy consumption in the sector: (1) lower energy costs for transporting goods in comparison with passengers, (2) the increase of energy cost as the speed rises and (3) the order of magnitude differences in energy cost between modes. Thus the lowest energy consumption is associated with modes engaged in the transport of bulky, homogenous material at low to moderate speeds while the highest is found in passenger and high value freight services operating at moderate to high speeds.

The variables which determine energy consumption in the sector may be grouped under the headings of technological considerations, operational and user behaviour characteristics and the broad structure and level of economic activity (Amman and Wilson 1980). Technologically, the key considerations are the thermodynamic efficiency of the vehicle's power plant and drive in converting fuel energy into propulsive thrust, the ratio of vehicle drag to total weight (frictional resistance) and the ratio of maximum potential payload to total vehicle weight.

Operationally, the important variables are load factor, quality and length of route and speed of operation. Energy is required to operate any vehicle whether empty or loaded. Consequently energy efficiency in terms of payload per kilometre travelled is greatest when the vehicle is carrying its maximum design load (i.e. when its load factor is 100 per cent). For a given mode and load factor the length of haul between stops is a significant influence on energy consumption. Thus short-haul intra-urban services (e.g. commuting and delivery) will be more energy-intensive than long-haul, intercity or international services whatever the mode used (UK, DOE 1976 and 1977b). Beyond the length of haul between stops is the question of the speed of operation within the general speed range of the vehicle concerned. Each vehicle has an engine design speed at which its energy consumption is minimized and operation above or below this speed will result in increased fuel consumption. Speed control is not only determined by the operator but also by route characteristics, traffic conditions and traffic control systems.

The demand for transportation services arises from the desire to move people and goods from place to place. The decision to choose a particular mode and rate to effect that exchange is influenced by a number of variables including cost, convenience, safety, speed and reliability. In the case of passenger transport the influence of speed and convenience (i.e. to higher energy using modes) increases with rising disposable income as may be seen from the global growth of automobile ownership and air travel. Once users have adopted a particular mode, habitual patterns develop which inhibit even inter-vehicle changes and seriously constrain inter-modal changes.

Broadly speaking, the demand for transportation services in any economy is determined by the general structure and level of economic activity, the spatial pattern of land use and the lifestyle of the population. Economies with large primary and secondary sectors actively engaged in trade will create a large demand for freight transportation and, if their population has a high disposable income, for passenger transport as well. In short, the internal and external level of spatial interaction and thus the demand for transportation services will be high. On the other hand, economies with large subsistence elements and low income will generate only a low to modest demand for usually short-haul passenger transport and long-distance goods movement will be required mainly by export oriented primary industries (Dunkerley et al. 1981).

A symbiotic relationship exists between the spatial distribution of land use and settlement and transportation services (Canada, Ministry of State for Urban Affairs 1977; Sewell and Foster 1980). For example, as commuter transportation services become available (e.g. streetcar, train, bus, auto) so the distance between residence and work place and personal service facilities extends. Once the land use pattern permitted by the availability of such services becomes established, the capital investment in that pattern becomes so great and the lifestyles dependent upon it become so entrenched, that it can change only very slowly. Consequently, what was initially the effect of transport availability becomes the cause of a continuing demand for it.

Summary

Before dealing with the geography of energy consumption it is necessary to develop a working knowledge of what might be called the 'anatomy of energy consumption'. This requires consideration of the components of consumption, the units in which they are quantified and the variables which shape them prior to overviewing representative consumption profiles.

Because of the multiplicity of energy consumers and the difficulty of collecting information from them, detailed data on energy consumption are less well developed than those for supply. Nevertheless there is a growing body of regularly published data which records consumption primarily by economic sector but, in special circumstances, by product system or function. These records use commodity (e.g. barrels or tonnes of crude oil), basic energy (e.g. BTU or joule) or derived energy units (e.g. energy efficiency, intensity and density).

The major variables which determine the amount and form of energy consumption are economic activity, price, population, and environmental conditions. Within the first it is the level of economic activity which has the greatest influence followed by the structure of the economy and the efficiency of the processes and technology in use. The real price of energy in relation to other inputs and the price differentials between energy sources and between regions all have a significant influence on the amount and kind of energy used. The number of people, their income, settlement pattern and lifestyle and the temperature regime, environmental regulations and spatial extent of the territory concerned complete the basic list of variables which decision makers (individual, corporate or government) consider in making energy choices.

Four general consumption profiles are presented based on economic sectors. Within them the product system and functional consumption approaches to energy consumption are illustrated. The domestic profile draws attention to the distinction between energy consumed in-house and beyond and illustrates the different consumption patterns which arise from variations in settlement pattern, income and lifestyle. The commercial profile illustrates the use of derived measures and documents the role which the functional structure of the sector and building characteristics play in determining the consumption pattern. The product system approach used in the industrial profile shows how energy needs differ between products and the relevance of process and technology in determining energy use. Finally, the transport profile illustrates, against a background of the overall level of economic activity as a determinant of the general amount of spatial interaction, the role played by mode choice and energy efficiency of the technology in use in determining consumption in this sector.

CHAPTER 3

Geography of energy consumption

The starting point of a study of the geography of energy consumption is an analysis of the spatial distribution of the components of consumption at a range of spatial scales (Fig. 3.1). In this chapter, after a background review of global trends, consumption will be broadly analysed by the standard international groupings and, within countries, by state (province) and urban areas.

Global trends

From 1945 to 1979, global consumption of commercial energy increased dramatically from 2200×10^6 to 8700×10^6 tonnes of coal equivalent (Fig. 3.2).

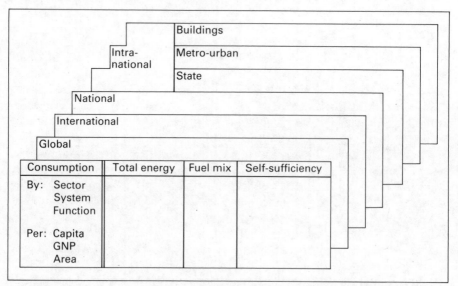

Figure 3.1 Energy consumption components at a hierarchy of scales

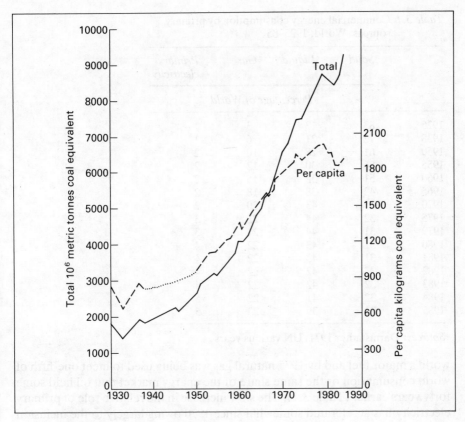

Figure 3.2 Commercial energy consumption: World, 1930–85
Sources: UN various years

When expressed in relation to population, this four-fold increase in total consumption resulted in the per capita index rising from approximately 600 to 2000 kg of coal equivalent. The strong upward trend which characterized these decades was brought about by the combined influences of population increase, economic expansion and rising incomes and, at least until the early seventies, a declining real cost of commercial energy.

The annual increases of 3–5 per cent in the global consumption of energy came to an abrupt end during 1980 when a decline of 0.5–1.0 per cent occurred until 1983. This sudden break from the norm of the previous thirty years was brought about by the combined influences of a very large increase in the price of energy and a world-wide economic recession. Although there have been individual years when consumption diminished or held steady, not since 1930 had there been such a sustained decline. An increasing trend of 2–3 per cent per year was re-established after 1983.

Fundamental changes in the mix of primary energy consumed accompanied the pre-1979 increase in consumption. By 1967 coal was replaced by oil as the

Table 3.1 Commercial energy consumption by primary
sources: World, 1925–85

	Solid	Liquid	Gas	Primary electricity
	Percentage of World			
1925	83	13	3	1
1938	72	21	6	1
1950	61	28	10	2
1955	56	30	12	2
1960	51	32	15	2
1965	42	38	18	2
1970	35	43	20	2
1975	32	44	21	3
1979	31	44	21	3
1980	31	43	22	4
1981	31	43	22	4
1982	32	42	22	4
1983	32	42	22	4
1984	32	41	22	5
1985	33	39	23	5

Sources: Darmstadter 1971; UN various years

world's major fuel and by 1975 natural gas was being used to meet one-fifth of
world consumption or the same share of the energy market that oil held some
forty years earlier (Table 3.1). The slow increase in the relative role of primary
electricity has accelerated somewhat since 1970 owing largely to the inclusion
of nuclear generation in this category (Table 3.2). In 1979 there was a

Table 3.2 Electricity consumption by primary sources: World, 1929–85

	Total $(10^9$ kWh$)$	Primary sources (percentage of total)		
		Hydro	Geothermal	Nuclear
1929	256	43	—	—
1950	872	38	<1	—
1960	2301	30	<1	1
1970	4956	23	<1	2
1976	6981	21	<1	7
1979	7966	22	<1	8
1980	8239	21	<1	8
1981	8357	21	<1	10
1982	8477	22	<1	10
1983	8825	22	<1	11
1984	9305	21	<1	13
1985	9675	21	<1	15

Sources: Darmstadter et al. 1971; UN various years

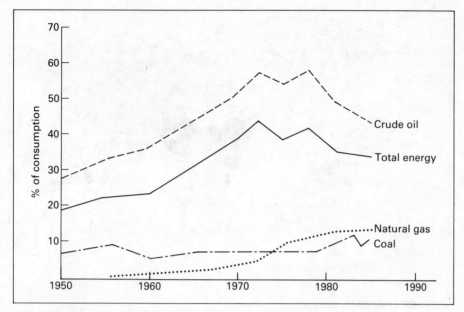

Figure 3.3 Fuel imports as a percentage of world consumption, 1950–85
Sources: UN various years

breakpoint in the mix of fuels used as well as for total consumption. Since then oil has decreased in importance, gas and primary electricity have continued to increase and the decline of coal has been halted. By the mid-1980s global energy consumption had entered a phase in which total demand had moderated and the fuel mix was in a transition moving towards the more balanced use of each of the primary sources of energy.

Over the approximately 60 years of the period under review the volume of foreign trade in energy has also changed considerably. During the long period when coal was dominant, imports did not contribute more than 20 per cent of the total energy consumed. However, as oil penetrated into almost all markets, reliance on imports increased substantially and, by 1972, over 40 per cent of the total energy consumed by the countries of the world was imported (Fig. 3.3). Since then the role of total imports has declined somewhat in response to a decline in imports of crude oil and in spite of the increased import of natural gas and coal.

International consumption

In the following review of energy consumption at the international scale, Developed and Developing Market Economies and Centrally Planned Economies are used as the major unit areas (Fig. 3.4). Sub-continental groupings and individual countries are referred to from time to time in order to describe some of the characteristics of the three broad groups more specifically.

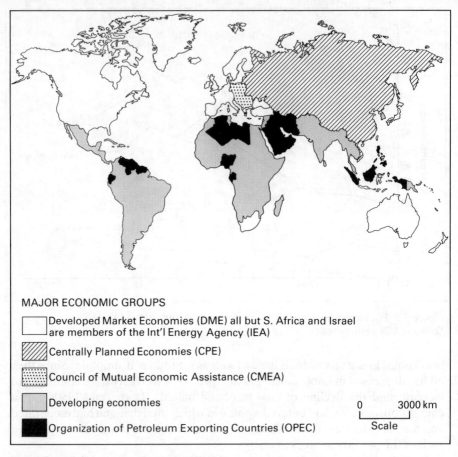

Figure 3.4 Major economic groups: World

Total energy

Table 3.3 depicts trends in the distribution of energy consumption among the major international groups since 1929. The Developed Market Economies have remained the major consumers of commercial energy in absolute terms throughout the period, but their relative importance has declined while that of the Centrally Planned Economies (CPE) has substantially increased. The decline in the former resulted from an approximately equal reduction in the proportion consumed in North America and Western Europe and despite an increase in that consumed by Japan. The rapid increase in the proportion consumed by the CPE has been largely the result of the growth of the USSR and, more recently, China as energy consumers. Although still only accounting for 12 per cent of global consumption, the rate of increase in the developing countries during the 1970s was higher than in either of the other two groups, the

50

Table 3.3 Commercial energy consumption: World and major economic groups, 1929–85

| | World | | Market Economies | | | | | | Centrally Planned Economies | | |
| | | | Developed | | | Developing | | | | | |
	Total*	Per** capita	Total	% of world	Per capita	Total	% of world	Per capita	Total	% of world	Per capita
1929	1799	866	1600	89	—	101	6	—	98	5	—
1955	3055	1143	2151	70	3504	185	6	157	720	24	816
1965	4992	1544	3129	63	4521	393	8	262	1470	29	1412
1970	6512	1781	4170	64	5739	518	8	302	1824	28	1500
1975	7529	1879	4447	59	5836	735	10	379	2347	31	1797
1979	8706	2019	4953	57	6317	933	11	437	2819	32	2027
1980	8566	1946	4798	56	—	1004	12	—	2745	32	—
1984	8889	1867	4597	52	—	1279	14	—	2980	34	—
1985	9130	1888	4874	53	—	1162	13	—	3094	34	—

* Total – 10^6 tonnes coal equivalent
** Per capita – kilograms coal equivalent
Sources: UN various years

most rapid increases being in the new industrial economies of countries such as Brazil and South Korea.

Fuel mix

The mix of commercial fuels used by the three international groups varies notably from the global aggregate (Table 3.4). The developing countries have the greatest dependence upon liquids which provide almost two-thirds of their total commercial fuel consumption. Indeed, more than 50 of the countries in this group rely on liquid fuels for more than 75 per cent of their requirements. By 1980, the dominance of hydrocarbons was further increased as natural gas replaced coal as the second major source. However, unlike liquid fuels which are widely used in all developing countries, consumption of natural gas is confined to a relatively small group of producers. Primary electricity (mainly hydro), although constituting only 6 per cent, contributes as much as in the Developed Economies.

The Centrally Planned Economies also have a distinctive fuel mix in that, as late as 1980, coal was used to meet 50 per cent of their requirements. Except for Cuba (which uses no coal) and the USSR which now relies on coal for less than one-third of its total requirements, the CPE of Eastern Europe and Asia use solid fuels for 65 per cent or more of their needs. Nevertheless, there has been a rapid increase in the use of liquid fuels in these countries (particularly in the decade 1960–70). The use of natural gas has been expanding even more rapidly as the result of the mobilization of the resources of the USSR and the development of pipelines to Eastern Europe.

In the Developed Market Economies, liquid fuels reached what may be their all time peak in 1979 when they supplied 50 per cent of total energy consumption. In Japan and several smaller countries, the proportion of oil was over 70 per cent but in the majority of economies in this group 45-50 per cent of the market is the norm for liquid fuels. Also in 1979, coal declined to its lowest relative importance since the beginning of the industrial revolution. The transition from solid to liquid fuels commenced in North America before World War II and only started on a large scale in Western Europe and Japan after 1945. As a result, the importance of coal at the end of the 1970s was a good deal less in Canada and the USA, than, for example, in the UK or West Germany. As in the other international groups, the role of natural gas has increased steadily, a development which also started in North America and only became a characteristic of Western Europe and Japan in the late 1960s with the introduction of marine transportation of liquefied natural gas. The role of primary electricity, although still small, is increasing as most of these countries added nuclear facilities.

By focusing on what is conventionally termed 'commercial' energy (fuels and electricity traded at a global and international scale), this discussion of the geography of energy consumption has ignored the role of 'non-commercial' or, perhaps more appropriately, 'traditional' energy sources. Even though the

Table 3.4 Energy consumption by percentage of primary source: major economic groups, 1929–85

| | Market Economies | | | | | | | | | | Centrally Planned Economies | | | | |
| | Developed | | | | | Developing | | | | | | | | | |
	Solid	Liquid	Gas	Elec.	Total	Solid	Liquid	Gas	Elec.	Total	Solid	Liquid	Gas	Elec.	Total
1929	84	12	3	1	100	59	38	2	1	100	86	13	1	—	100
1955	49	33	16	2	100	27	64	6	2	100	82	14	3	1	100
1960	40	37	20	3	100	23	66	8	2	100	79	14	6	1	100
1965	34	43	20	3	100	22	64	11	3	100	66	19	14	1	100
1970	26	48	23	3	100	18	65	13	4	100	60	24	15	1	100
1979	23	50	23	4	100	15	67	13	4	100	52	27	20	2	100
1980	24	48	24	4	100	15	64	16	5	100	50	28	21	2	100
1984	26	46	22	6	100	20	60	15	5	100	48	25	25	2	100
1985	26	46	22	6	100	25	54	15	6	100	48	24	26	2	100

Sources: UN various years

Table 3.5 Traditional energy sources as percentage of total energy requirement: selected countries, 1967 and 1982

	Total Energy Requirement (TER)		Traditional energy sources as percentage of TER	
	1967 *10^6 m.t.o.e*	*1982* *Terajoules*	*1967*	*1982*
Argentina	26.8	1843	7.9	5.6
Brazil	54.3	6111	42.5	32.6
Colombia	14.2	1048	40.3	15.8
Egypt	6.4	851	2.2	3.9
India	87.7	7679	28.3	29.5
Indonesia	31.7	2309	70.6	49.2
Jamaica	1.7	109	8.8	4.6
Kenya	3.5	347	72.1	77.5
S. Korea	13.6	1936	15.6	3.5
Nigeria	14.2	1291	87.5	57.4
Thailand	4.0	1007	47.4	46.9
Venezuela	17.3	1738	9.4	0.9

Sources: IEA 1979; UN 1985

modernizing and urban sectors of these economies are making the transition to commercial fuels, the majority of the population has very little access to them and relies heavily upon fuels gleaned from local biomass and waste. The relative role of commercial and traditional fuels in the total energy consumption of developing countries as a whole, however, is difficult to assess because of the fragmentary nature of the data. For individual countries, an indication of the relationship and the way it changed between 1967 and 1982 is shown in Table 3.5. While the proportion of the total energy requirement met by non-commercial sources ranges from almost 90 per cent to less than 1 per cent among the countries listed, most countries exhibit a downward trend over the decade.

Self-sufficiency

Prior to 1925 there was a considerable degree of energy self-sufficiency in the major economies of the world. Coal production was relatively widespread and such imbalances between consumption and production as occurred were accommodated largely by intra-regional trade. After 1925, this pattern began to change as Western Europe's small but increasing liquid fuel requirements were met by imports. Elsewhere in the developed world at this time, the production/consumption ratio was above unity and the developing economies were already divided into a relatively small group with a substantial surplus of energy and a large, heavily populated group with a deficit (Table 3.6). By 1980, a clear division had developed between energy deficit and energy

Table 3.6 Production/consumption ratios of commercial energy: major economic groups and regions, 1925–85

	1925	*1955*	*1980*	*1985*
Market Economies				
Developed	104	92	71	76
W. Europe	92	83	52	61
N. America	104	100	89	91
Japan	108	76	10	11
Oceania	103	67	117	159
Developing	134	289	256	218
Energy export countries	332	1517	634	396
Other	60	72	74	76
Centrally Planned Economies	109	105	114	112
USSR	107	100	131	127
E. Europe	118	106	76	76
Asia	95	99	109	111

$$\text{Production/consumption ratio} = \frac{\text{Production}}{\text{Consumption}} \times 100$$

Sources: Darmstadter *et al*. 1971; UN various years

surplus economies. All the major sub-groups of the Developed Market Economies except Oceania are now in an energy deficit position as is Eastern Europe in the CPE group. Conversely, there has emerged a relatively small group of energy surplus countries including those developing nations rich in hydrocarbons and the USSR.

Table 3.7 illustrates these broad trends by reference to a selection of individual countries. In Western Europe there is a growing dichotomy between those sharing in the development of the hydrocarbon resources of the North Sea (Netherlands, UK and Norway) which have a surplus, and the others with a large deficit (e.g. France and West Germany). In North America there is a complementarity between Canada and the USA. As the US demand for oil and gas exceeded domestic supply, Canada became a supplier of oil, then gas and, increasingly, of electricity. Japan is the most energy-deficient of all the developed economies, having to import very large quantities of oil, gas and coal.

Among the developing economies, the first four countries in Table 3.7 are representative of the hydrocarbon export group of which three are members of the Organization of Petroleum Exporting Countries (OPEC). Brazil and South Korea represent the newly industrializing countries which are able to provide less than half of their own fuel requirements. In the CPE, both the USSR and China consume less than they produce in all fuel categories while the Eastern European countries have a deficit in all categories except Poland which has a surplus of coal.

Table 3.7 Production/consumption ratios of commercial energy: selected
countries, 1985

		Total commercial energy	Solid fuels	Crude oil	Natural gas
Market Economies — Developed	UK	107	63	150	74
	Netherlands	115	—	53	205
	France	29	50	4	23
	W. Germany	45	98	4	31
	USA	91	114	73	96
	Canada	125	128	120	138
	Japan	9	15	—	6
	Australia	153	230	92	100
	New Zealand	78	118	29	100
Developing	Saudi Arabia	893	0	930	100
	Algeria	664	0	967	473
	Nigeria	499	0	754	100
	Mexico	200	95	242	106
	Brazil	72	48	65	100
	S. Korea	91	98	—	0
	India	99	103	87	100
	Bangladesh	63	—	5	100
Centrally Planned Economies	USSR	128	103	177	113
	Poland	104	121	2	44
	Rumania	90	78	93	96
	Hungary	56	73	32	66
	China	109	101	158	100

0 = No production or consumption
— = Consumption but no production
Source: UN 1987

Consumption by sector

Comparison of energy consumption by economic sector in terms of the three
major international groups is not possible because of the absence of data at this
scale. There are, however, good data for the OECD countries (OECD 1983)
and a start has been made for the developing countries under the auspices of the
International Energy Agency (IEA 1979) and by Resources for the Future
(Dunkerley 1980). Consequently, individual countries rather than international groupings will be used as the unit areas in this section.

At the national scale, there are several reasons why sectoral consumption
data have to be interpreted with care when used for comparative purposes.

First, and despite the standardization attempts of agencies such as IEA and RFF, definitional and measurement variations may persist between countries and, because of the difficulties of recording energy use, the 'not included elsewhere' or 'other use' categories are often large (particularly for developing countries). Second, and again particularly in the developing countries, it is important to determine if it is consumption of total energy (i.e. commercial and traditional) or just commercial energy that is recorded. Third, it must be clear whether the energy shown as being consumed in each sector is primary (i.e. before conversion or transport losses are considered) or secondary energy such as thermal electricity and petroleum products. This distinction is recognized by referring to the former as total energy requirement (TER) and the latter as total final consumption (TFC). The differing impressions of sectoral consumption that can arise as a consequence of some of these data variations are illustrated in Tables 3.8 and 3.9. Commercial energy and total final consumption are used in the remainder of this section.

Examination of the available data on consumption by sector for individual countries reveals two major classes (Table 3.10). In the first, industry is the dominant sector and, in the second, transportation. Within the first group, there is a further division on the basis of the sector which is the second largest consumer. Sub-group A, in which industry and transportation are the major consuming sectors, includes countries in which primary processing industries play a large part in the industrial structure and the sheer spatial extent of the national territory requires long distance transportation systems. In sub-group B, industry is followed by residential as the second major consuming sector. This profile is representative of most West European countries in which the industrial structure is much more varied, incomes are high and transportation systems are compact.

Table 3.8 Comparison of sectoral consumption of total primary energy and commercial energy only: selected countries, 1977

		Energy industry	Industry	Transport	Other (agric; res/comm; non-energy)	Not included elsewhere
Brazil	Total	21.0	22.8	19.9	34.2	2.1
	Commercial only	28.7	25.7	27.1	7.6	10.9
India	Total	20.3	31.3	13.3	34.0	1.1
	Commercial only	26.8	39.6	17.6	12.8	3.2
Kenya	Total	10.8	7.8	13.3	67.7	0.4
	Commercial only	28.4	17.1	35.1	9.3	10.1

Source: IEA 1979

Table 3.9 Total energy requirements and total final consumption by sector: selected countries, 1981

		Energy industry	Industry	Transport	Residential	Other
		(Percentage of consumption)				
USA	TER	25.6	22.4	24.4	19.0	8.6
	TFC	—	30.1	32.8	25.6	11.5
Canada	TER	33.0	19.9	19.4	13.9	13.8
	TFC	—	29.7	29.0	20.7	20.6
UK	TER	30.7	27.2	14.8	16.8	10.5
	TFC	—	39.1	21.3	24.2	15.4
Australia	TER	36.9	29.3	21.0	7.0	5.8
	TFC	—	46.5	33.3	11.1	9.1
New Zealand	TER	34.9	22.9	21.1	11.0	10.1
	TFC	—	35.2	32.4	16.9	15.5

TER = Total energy requirements
TFC = Total final consumption
Source: IEA 1983a

The second group consists of countries in which the transportation sector is of greater relative importance than industry. The dominance of transportation in the US may be attributed to the sheer magnitude of the tonne/kilometres generated by a highly developed, complex economy set in an extensive area with a large, affluent population whose stock of personal transportation

Table 3.10 Percentage of total final consumption of commercial energy by sector: selected countries

			Industry	Transport	Commercial	Residential	Other*
I	A	Mexico	54	25	1	4	16
		Australia	41	37	6	8	8
		Canada	37	26	14	16	7
	B	Japan	52	19	2	22	5
		UK	32	24	12	27	5
II		USA	28	34	11	17	10
		Brazil	36	38	2	3	21
		Argentina	22	45	4	16	13
		Thailand	28	44	6	4	18
		Kenya	24	55	4	7	10

* Other – includes agriculture, non-energy use and energy included in TFC but not distributed sectorally
Sources: IEA 1979, 1983

equipment exceeds that in any other country. In Kenya, on the other hand, the dominance of transportation is more likely to be a residual characteristic arising because commercial energy consumption in other sectors is small, reflecting low income and little manufacturing.

Intranational consumption

The spatial variation of consumption profiles varies almost as widely within countries as it does among them. This section illustrates these variations at two scales using states (provinces) and urban centres as the unit areas.

Profiles by state

The data for Tables 3.11–3.13 are drawn from publications of national energy organizations in Australia, Canada and the USA. The units, stage of consumption and terminology used vary from country to country, reflecting the different conventions used by each and illustrating the difficulty of assembling data which permit comparisons between countries (Dunkerley 1977, 1980). It should also be noted that the data are not comparable to those used for the same three countries in Table 3.9. The regional differences illustrated by the table for each country reflect the complex relationship that exists between fuel mix and sectoral consumption. As a first approximation, however, variations in the mix of primary energy consumed are generally the result of the relative ease of access to the respective primary energy sources, while the sectoral consumption is a function of economic structure.

Australia The data for Australia refer basically to primary energy so that electricity is not distributed among the sectors but the fuel used for electricity generation is included in a sector labelled 'utilities'. Two other features of the data should also be noted: petroleum products (secondary energy) are included and distributed among the sectors and consumption in the domestic sector is included with commercial.

In the national fuel mix, coal and liquid fuels are co-dominant, followed by natural gas and biomass (wood and bagasse) but only a small contribution from hydro and no nuclear (Table 3.11). Among the states, Western Australia has the most extreme dominance of oil while in New South Wales and Victoria coal replaces oil as the major fuel used. Cane-growing Queensland and relatively well-forested Tasmania consume significant quantities of biomass though it is also used sufficiently to be recorded in the other states. Solar energy contributes small but recorded amounts in three states.

Sectorally, the consumption profiles show manufacturing to be the major consuming sector in all states with utilities and transportation vying with one another for second place. The domestic/commercial sector accounts for remarkably little of the total consumption in all states but Victoria and Tasmania.

59

Table 3.11 Energy consumption profiles of states of Australia, 1983

	Australia	New South Wales	Victoria	Queensland	South Australia	West Australia	Tasmania	Northern Territories
Primary fuels consumed (PJ)	3110	997	895	539	253	305	81	41
Sources – %								
Solid	39	52	38	41	17	22	10	<1
Liquid	39	36	33	42	39	62	47	100
Gas	15	8	25	3	42	13	—	—
Hydro	2	1	1	<1	—	—	34	—
Biomass	5	3	3	14	2	3	9	—
Solar	<1	<1	—	—	<1	<1	—	—
Sectors – %								
Industrial	31	30	23	34	29	31	48	42
Transport	28	28	24	26	27	32	23	27
Utilities	28	28	36	24	26	20	5	22
Domestic-commercial	11	13	16	11	14	13	22	12
Agriculture	<1	1	1	5	4	4	2	2

Source: Australia, Department of Resources and Energy 1984

Table 3.12 Energy consumption profiles of provinces of Canada, 1985

	Canada	Atlantic Provinces	Quebec	Ontario	Manitoba	Saskatchewan	Alberta	British Columbia	Yukon NWT
Production (PJ)	9931	272	493	371	112	699	6655	1266	62
Export (1) (PJ)	3202	96	77	263	1059	242	4155	1301	28
Import (1) (PJ)	1081	300	838	2300	923	—	—	727	—
Total energy availability (PJ)	7810	476	1254	2781	142	378	2018	732	18
Sources – %									
Solid	14	18	1	19	3	35	16	1	—
Liquid	39	70	47	40	43	20	32	43	39
Gas	32	—	14	29	31	43	50	33	50
Hydro/nuclear	15	12	38	12	23	2	2	23	11
Final consumption (2) (PJ)	5846	402	1226	2149	231	295	869	647	28
Sectors – %									
Industrial	33	27	37	35	17	23	30	35	30
Transport	28	32	26	27	35	32	26	30	22
Agriculture	3	2	2	2	6	12	5	1	—
Residential	19	19	20	19	20	17	20	17	11
Commercial	17	20	15	17	22	16	19	17	37

(1) For provinces includes inter-provincial transfer
(2) Primary and secondary
Source: Statistics Canada 1986

Canada The data for Canada differ from those for Australia. They include information on self-sufficiency, they distinguish between total energy availability and final consumption, and electricity is distributed among the sectors (Table 3.12). Within Canada, 58 per cent of national consumption occurs in central Canada (Ontario and Quebec) and over 80 per cent of current energy production is in western Canada, particularly in Alberta.

Liquid fuels dominate consumption from Manitoba eastwards with gas increasing in importance from Quebec westwards, reaching a maximum in Alberta. In Quebec, Manitoba and British Columbia primary electricity (hydro) provides a major proportion of consumption needs.

In Central Canada (Quebec and Ontario) and the two westernmost provinces (Alberta and British Columbia) the industrial sector is the major consumer with transportation dominant in the other provinces. Relative consumption in the residential sector is remarkably consistent across Canada (except for the Yukon and Northwest Territories). If residential and commercial were combined, as in the Australian data, the joint sector would be the major consumer in all provinces but Quebec and British Columbia.

USA Each of the state fuel mix profiles shown in Table 3.13 is broadly representative of a group of surrounding states and thus illustrates the regional variation of fuel mix in the US. The dominance of oil followed by gas in New York is found in all the east coast states although a few use coal (rather than gas) as the second major fuel. With little or no energy production and a large population these states collectively constitute the major energy deficit area of

Table 3.13 Energy consumption profiles of selected states of the USA, 1985

	USA	California	New York	Penn-sylvania	Texas	Washington
Total consumption:						
10^{12} BTU	74,023	6,358	3,372	3,271	8,924	1,683
Per capita:						
10^6 BTU	310	241	190	358	545	382
Sources – %						
Solid	24	1	9	43	13	6
Liquid	41	50	47	37	40	34
Gas	24	30	23	20	47	8
Nuclear	6	3	8	9	—	5
Hydro	5	6	13	1	—	47
Sectors – %						
Industrial	36	27	21	39	55	33
Transport	27	38	25	23	23	25
Residential	21	19	27	24	12	24
Commercial	16	16	27	14	10	18

Source: US, DOE 1987

the country which has to rely on imports mainly from the Gulf states, Canada and offshore. Pennsylvania, the only state of the five selected with coal as the major fuel, has a mix typical of the north-central group of states. Texas, on the other hand, in which gas is the dominant fuel, is representative of several of the large hydrocarbon producing states in the south. Washington is the only state in which hydro is dominant, although it plays a major role in the adjoining states of the Pacific Northwest.

Four of the five states included in Table 3.13 have basically the same sectoral consumption profile with industry as the leading consumer followed by transportation, residential and commercial in that order, except for Pennsylvania in which the residential sector exceeds transportation by a small margin. Among these four, however, the precise role of each of the sectors varies. Texas and Pennsylvania have the most pronounced dominance of industry and the least consumption by the commercial sector while New York, although conforming to the general pattern has only a small range between the sectors. In California, with its highly mobile population and few energy intensive primary processing industries, the transportation sector replaces industry as the major consuming sector.

Metropolitan–urban consumption

Analysis at international and national scales provides only a very coarse-grain impression of the spatial distribution of energy consumption. Even when the resolution is increased by considering data at the intranational scale by using states (provinces), the essence of the distribution is still not revealed because the real focus of energy consumption is the city. With the exception of the energy consumed in inter-city transport and in primary economic activities such as agriculture or mining, the consumption of energy is essentially a function of the level of economic activity and population density both of which reach a maximum in the cities of the world.

Data for urban energy consumption, however, are not regularly published in any country. Rather, the quantitative information available is contained in special studies, many of which have been undertaken in a climatological context and are aimed at heat island or atmospheric quality questions (Kalma et al. 1978; Oke 1973). In these studies the energy consumption statistics are usually derived from other data. In many, national or state data are disaggregated by means of applying general indices and deriving urban totals from them (Kalma 1976). Statistics generated in this way are very sensitive to the values used to determine the indices and at best yield derived rather than actual data. Urban scale studies of consumption by particular sectors or activities often use one-time sample surveys of actual energy use. Such surveys yield more accurate data than the disaggregation approach but have the disadvantage of being limited in scope to the sector being sampled.

To illustrate the substantive aspects of urban energy consumption the first part of this section is based upon one of the best documented, although dated, studies available. This study of the New York region (RPA/RFF 1974; Darmstadter 1975) was specifically designed to study the energy consumption characteristics of a large and complex urban region. Few other such studies are available (see, however, Kalma 1976; Newcombe *et al.* 1978) and none illustrates better the changing structure of consumption as the city core is approached. The New York Region extends over parts of three states (Connecticut, New York and New Jersey) and may be subdivided into the suburbs, city and inner city (Table 3.14).

The most notable characteristic is the sheer magnitude of energy consumed. In 1970 the gross energy consumption of the New York region was larger than that consumed by all but seven of the countries of the world (excluding the USA) and that of New York City itself more than any developing country and even some Western European states. This is a particularly dramatic illustration of the energy intensity of large, modern cities and, since primary energy production rarely occurs within their boundaries, the necessity of very large-scale energy supply systems to meet their requirements.

The mix of energy forms which make up the total net consumption (i.e. consumption excluding the energy used to generate electricity) differs little from one scale to the other except in the inner city where the role of electricity increases significantly. Sectorally, transportation is the major consumer in the

Table 3.14 Energy consumption in the New York region, 1970

	New York region	New York suburbs	New York city	Manhattan
Population (10^3)	19,756	11,860	7,896	1,539
Gross energy consumption (10^{12} BTU)	4,186	2,782	1,404	322
Per capita (10^6 BTU)	212	235	178	209
Net energy consumption (10^{12} BTU)	3,398	2,301	1,097	252
Per capita (10^6 BTU)	172	194	139	173
Net energy consumption		*Percentage of total*		
By energy form				
Electricity	9	8	9	17
Gas	15	16	14	8
Coal	2	2	1	—
Liquid	74	74	76	76
By sector				
Residential	32	28	38	33
Commercial	26	24	28	44
Industrial	9	11	6	7
Transportation	34	37	28	16

Source: Regional Plan Association, Resources for the Future 1974

region and suburbs, residential in the city as a whole and commercial in the inner city. This distribution of sectoral consumption reflects the well-known differences in economic structure, land use and public transportation availability that occur between the outer and inner city.

Summary

After a brief review of global trends, the geography of energy consumption is described in this chapter at three scales. Internationally the unit areas are the major economic groups and regions and intranationally they are states (provinces) and urban regions. Total energy, mix of sources, sectoral use, trade and self-sufficiency are the major components of energy consumption considered.

From 1945 to 1979 global commercial energy consumption increased annually at rates varying between 3 and 5 per cent, but, between 1980 and 1983, the trend was reversed and decreases of 0.5 to 1.0 per cent were recorded. After 1983 annual increases have been established but at rates between 2 and 3 per cent. Spatially, since 1945, the relative importance of the Developed Market Economies as energy consumers has declined and that of the Centrally Planned Economies and, especially, the Developing Economies has increased. Globally, the use of oil rose steadily to reach a maximum of 44 per cent of total commercial energy consumed in the late 1970s, declining thereafter as coal and natural gas were substituted. The shift to oil was relatively greatest in the Developing Economies where, in 1979, oil constituted two-thirds of their commercial energy consumption. In the Centrally Planned Economies, although coal continues to be the major fuel used, oil consumption increased providing a maximum of 28 per cent of all energy consumed in 1980. Since then the proportion of oil has declined as the use of natural gas has increased.

After World War II, first Western Europe, then Japan, Eastern Europe and the USA have become increasingly large energy importers giving rise to world trade becoming dominated by energy commodities. The Organization of Petroleum Exporting Countries emerged as the major supplier of oil to these markets followed by the USSR (oil and gas) and a group of coal exporting countries which includes the USA (though an importer of oil and gas), Australia, Canada and others.

Data on the consumption of energy within countries are readily available for the Developed Market Economies and some of the Developing Economies. When the unit areas are states (provinces) rather than whole countries, major external differences are revealed in per capita consumption, energy sources used and sectoral requirements. At the city and urban region level, primary data are rarely available despite the central importance of these units as nodes of energy consumption. In most available urban based studies, data are derived by disaggregating state or national information by applying indices such as consumption per capita, worker or unit of production. The example presented in this chapter, although prepared in the early 1970s, represents the most comprehensive attempt to document energy consumption at the urban scale using primary data. It clearly illustrates the magnitude and concentration of urban energy consumption and suggests directions which will become increasingly common if Third World urbanization continues to increase as expected.

This chapter has illustrated the spatial variation of some of the important components of energy consumption by reference to unit areas of different sizes ranging in extent from millions of square kilometres (international groups) to less than ten square kilometres (city centres). At all of these scales, the spatial concentration of commercial energy consumption is the most notable feature. Internationally less than a dozen countries consume almost 75 per cent of the world's commercial energy; within individual countries, the major urban centres consume the majority of the energy used and, within those centres, particular districts and even large buildings and factory complexes constitute nodes of very high consumption density.

Two important consequences follow from this spatial concentration:

1. in order to meet the sheer magnitude of demand, very large scale and reliable supply systems must be available;
2. the high density of combustion residuals can lead to acute environmental deterioration at local and regional scales and long-term degradation over larger areas.

Beyond the effects of the concentration of consumption, the spatial variation of fuel mix, sectoral consumption and self-sufficiency sketched in this chapter constitutes the kaleidoscope of reality with which energy managers have to contend. The variety of consumption characteristics found within countries often makes concerted national policy difficult to implement. Similarly, at the international level, the variety of national profiles constrains international cooperation. However, to dwell exclusively upon spatial variety is to lose sight of the similarities which exist and which can form the basis of the common interests required to produce agreement and cooperation.

CHAPTER 4

Commercial energy supply systems

In order to understand the geography of the energy supply industry (Chapter 5), it is necessary to be familiar with the basic structure and terminology of the major sectors. In this chapter the general characteristics of the commercial fuel industries (coal, oil, natural gas and uranium), a selection of renewable resource-based systems and the electricity industry are reviewed, making use of a modified version of the structure of energy systems introduced in Chapter 1 and summarized in Table 4.1.

Table 4.1 Framework of headings for review of energy supply systems

Functional components – the energy supply chain
 Resource base
  Characteristics and classification
  Exploration and delineation
 Production
 Processing
 Transportation and storage
Spatial characteristics
Environmental aspects
Organizational structure

The general characteristics of the supply chain for each of the major energy sources are shown diagrammatically at the beginning of the respective sections and the major links are systematically reviewed. The major spatial, environmental and organizational characteristics are then described. This approach provides a broad framework within which to compare the structure of the individual energy supply sectors and a basis for a review of their respective geographies.

67

Commercial energy supply systems

Coal

Coal is a combustible substance consisting of fossilized plant material formed from the progressive alteration of woody vegetation by high pressures and temperatures. In comparison with other fossil fuels, coal is relatively widely distributed at the global scale and, in total, constitutes by far the largest non-renewable source of energy available to mankind.

Functional components

The functional components of the coal industry are illustrated in Figure 4.1 which displays a chain of facilities from primary production, preparation, processing to transport. Each of the links in the chain requires substantial capital investment (Wilson 1980) and long lead times to construct.

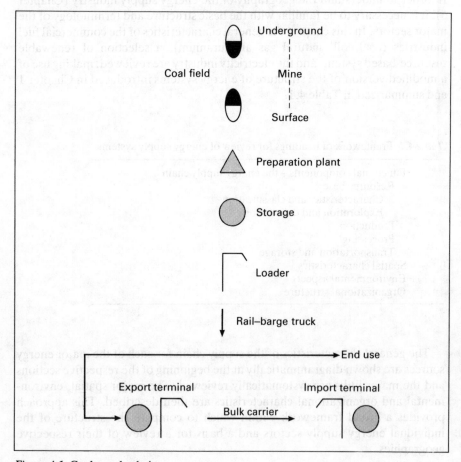

Figure 4.1 Coal supply chain

Resource base

The value of coal as a fuel or raw material depends largely upon its chemical and physical characteristics which, in turn, result from the qualities of the initial vegetation and the temperature and pressure conditions to which it has been subjected.

Characteristics and classification   The characteristics brought about by the formative conditions are referred to as 'coalification' and result in four major classes of coal (Table 4.2) each of which has different chemical and heating properties. Among the chemical characteristics, the amounts of carbon and volatile matter are particularly important measures of quality and the amounts of sulphur and ash lead to important waste product considerations at the point of combustion.

The physical characteristics of coal include such qualities as hardness, strength and agglomerative behaviour as well as the structural aspects (e.g. thickness, regularity) of the coal bed. As with other sedimentary strata, coal beds vary in depth below the surface, thickness and continuity, characteristics which play an important part in determining the manner and cost of production. Generally speaking, thick (e.g. 15–30 m, 50–100 ft), continuous seams that are close to the surface (e.g. 15–30 m, 50–100 ft) are the cheapest to mine. Agglomerating coals can be converted into coke and in many parts of the world are referred to as 'metallurgical' coal while non-agglomerating coals are put into the general category of 'thermal' ('steam') coal.

'Exploration for discovery' is much less a concern for coal than for hydrocarbons because, at least in the northern hemisphere, the existence of the major coal basins has been known for a long time. However, with the resurgence of interest in coal as an alternative to oil or gas, there has been a considerable increase in more detailed appraisals not only in the older, well developed coal areas of the world but in relatively new areas such as north-eastern Australia, western Canada, western USA and Siberia, as well as in the developing world. Because coal beds extend relatively continuously and laterally over distance and, in comparison to hydrocarbons, occur at shallow depths, delineation drilling is both less uncertain and less costly.

Production

Coal is produced in three different ways: underground mining, surface (strip, open pit or cast) mining and augering. Production data usually distinguish between gross mine output and marketable production after sorting and washing.

Underground mining is the oldest commercial means of coal production. Initially very labour intensive, underground mines have become increasingly capital intensive, the degree and type of mechanization being directly related to the thickness and continuity of the coal seams.

Table 4.2 Analyses of representative coal types in the US

Rank	State	Percentage				Percentage*					Heating value BTU/lb
		Moisture	Volatile matter	Fixed carbon	Ash	S	H	C	N	O	
Anthracite	Pa.	4.4	4.8	18.8	9.0	0.6	3.4	79.8	1.0	6.2	13,130
Bituminous coal											
Low-volatile	Md.	2.3	19.6	65.8	12.3	3.1	4.5	74.5	1.4	4.2	13,220
Medium-volatile	Ala.	3.1	23.4	63.6	9.9	0.8	4.9	76.7	1.5	6.2	13,530
High-volatile A	Ky.	3.2	36.8	56.4	3.6	0.6	5.6	79.4	1.6	9.2	14,090
Sub-bituminous coal	Wyo.	22.2	32.2	40.3	4.3	0.5	6.9	53.9	1.0	33.4	9,610
Lignite	N. Dak.	36.8	27.8	30.2	5.2	0.4	6.9	41.2	0.7	45.6	6,960

* S = Sulphur, H = Hydrogen, C = Carbon, N = Nitrogen, O = Oxygen
Source: Parker, S. P. 1981

Surface mining is practical where the coal beds occur at depths up to approximately 70 metres (200 ft) and thus are accessible from the surface. One method, commonly referred to as strip mining, involves the sequential removal of the overburden in strips and its deposition in a series of parallel ridges. In the other major method of surface mining, open pit, the overburden is removed from the pit and dumped on adjacent land. The total area involved in this type of operation may be as great as several square kilometres. The productive capacity of surface operations is generally much greater than underground mines and some of the largest have an annual capacity in excess of 50 million tons a year (Shabad 1984). In such mines production and recovery rates are high and costs per tonne are low.

Augering is a special method of extraction, used mainly in the USA, to produce coal which is not accessible by other means. Single or multiple augers (which are like very large carpenter's drills) usually draw out only about half of the coal in a given seam.

The cost advantages of surface mines have encouraged their development wherever physically possible and the proportion of world coal produced in this way has increased rapidly. Planned production from large, shallow coal fields in Siberia, western North America, and north-eastern Australia may accelerate this trend. Nevertheless, the majority of production from the older, developed fields will continue to be from underground mines although the number of mines in a given coal basin will decrease as comprehensive coal-*field* management practices are introduced.

Processing

After coal is extracted it may be shipped from the mine just as it is produced or it may be beneficiated. Such preliminary processing (referred to as coal preparation) is necessary when the run-of-the-mine quality falls below that specified by the customer and/or the amount of non-combustible material adds significantly to transport cost. Mine-mouth preparation plants may carry out sizing and flotation grading (to remove the free non-combustible material) and dewatering (mechanical) and drying (thermal).

More advanced stages of processing transform or convert coal into a large number of other energy and raw material products. The most important of these are briquettes (patent fuel) and coke, synthetic gas and oil (synfuel) and chemical feedstocks. The 'carbonization' of coal results in either *coke* or *char*. The latter, produced in low-temperature ovens or as a residual in some coal gasification processes, may be marketed directly as fuel or crushed, mixed with asphalt, compacted and heated and sold as *briquettes*. The production of metallurgical coke by the distillation of coking coal is a very much larger scale and more widely distributed form of coal processing (344 million tonnes compared with 97 million tonnes of *briquettes* in 1984). The majority of metallurgical coke is produced in coke ovens, usually located immediately adjacent to the blast furnaces of an integrated iron and steel plant. Valuable

Table 4.3 Heat value of manufactured gas

Rank	Heat value	
	BTU/ft^3	kJ/m^3
Low	150	4.5
Intermediate	500	15.0
High (Substitute or synthetic natural gas)	1000	30.0

Source: Parker, S. P. 1981

by-products include low-intermediate gas and liquids which may be used as fuel or feedstock for the chemical industry.

Coal (and coke or char) can be converted into gas. The chemical composition and heating value of the gas is determined by the *gasification* process used and the properties of the feed. The uses and transportability of manufactured gas are largely a function of its heat value (Table 4.3). Gases with low and intermediate heat values are now mainly used for in-plant industrial purposes. Coal-based manufactured gas of intermediate rank was distributed by the town gas systems of the industrial world until natural gas became available. Now gasification of coal is being considered as a means of producing a substitute for natural gas, a way of dealing with the emission problems of high sulphur-coal combustion, combined-cycle electricity production and in conjunction with some coal liquefaction processes.

In many respects the most elaborate stage of coal processing is *liquefaction*. The production of petroleum liquids by direct or indirect (gas to liquid) coal liquefaction methods is significantly more costly than producing them from conventional hydrocarbon sources (even when crude oil was priced at US $34 a barrel as in the early 1980s). South Africa, with very low coal production costs, is the only country with a large-scale coal liquefaction industry. Elsewhere, coal could be converted into liquid or gaseous hydrocarbons. First and second generation technologies are operational at a commercial scale, but third generation processes capable of very large-scale, low-cost production are still in the development stage. When the cost differential between the 'manufac-tured' and 'natural' products is decreased, coal based synfuels could become current resources.

Transportation and storage

Coal is a bulky material (0.8–1.0 tonne/m^3) produced in very large quantities that often have to be transported long distances from mines to markets. The transportation systems required for this movement include bulk loading and unloading terminals (each with space for open or contained storage) and a trunk transporter consisting of rail, water-borne or, in a few cases to date, pipeline carriers (Fig. 4.2).

For medium to long distance intra-continental transportation, coal is carried by unit-trains, vessels on inland waterways and, in a few instances in the USA and USSR, by slurry pipelines. The use of trucks and conveyor belts, though highly developed in some areas, is limited to short hauls. In North America a standard unit-train consists of 100 cars, each with a capacity of 100 tonnes (i.e., a 10,000 tonne load). Such trains require a heavy duty track and rail bed and, if the potential economies are to be realized, a control system that permits a relatively uninterrupted passage.

Slurry pipelines are designed to transport ground-up coal suspended in water. Although the only commercial lines that have been built are less than 15 km (10 miles) in the USSR, 150 and 450 km (100 and 275 miles) in the USA, lines of 1600–3200 km (1000–2000 miles) with capacities of 25–50 million tons are said to be in an advanced stage of design in the USA and USSR (Wasp 1983). Future construction of slurry pipelines is only likely under conditions which guarantee high load factors and where the point of origin is well endowed with water. This latter constraint arises from the fact that not only are very large quantities of water required for this mode of transportation but the water is usually not returned to the original watershed.

Inter-continental transport of coal, by definition, involves vessels, an increasing proportion of which consists of large (50–175,000 DWT) bulk carriers. Such vessels require deep-water terminals with high capacity equipment for loading, unloading and mixing as well as extensive areas for marshalling yards and storage facilities. The largest export complex is the Hampton Roads, Norfolk facility on the east coast of the USA, with a capacity in excess of 40 million tonnes per year. Others such as Richard's Bay, South Africa and Hay Point and Waratak in eastern Australia were able in the mid-1980s to handle approximately 20 million tonnes per year. The largest import terminals are in Japan and Western Europe in both of which new and expanded facilities are planned.

Spatial characteristics

The terminology used to describe the spatial distribution of coal ranges from basins and fields to individually identified seams, the thickness and horizontal and vertical extent of which has to be mapped in detail in order to design production facilities. Coal deposits occur in spatial association with sediments as old as Devonian and as recent as Tertiary. Generally speaking the older the occurrences the higher the rank of associated coal. Topographically, coal fields may be associated with flat or gently rolling surface terrain as well as highly accidented, mountainous land forms. In the latter, as a consequence of more active metamorphosis, geologically recent coals may be of a higher rank than would be expected and, as a consequence of the folding and faulting associated with mountainous regions, are often discontinuously distributed.

From the point of view of spatial interaction, one of the most important variables affecting the transferability of coal is its heat content. Coals with low heat value and/or high ash content are too expensive to be moved very far and thus have to be used close to the mine or converted into a more transferable form. However, most bituminous coal and even high rank sub-bituminous coal have sufficient heat value and, if necessary, can be beneficiated to a quality to permit them to be transported over long distances by efficient bulk handling carrier systems.

Environmental aspects

A variety of environmental disturbances and residuals are produced along the coal supply chain. Although the effects of coal combustion may extend over regional and even continental scales (acid rain), most other environmental effects are local in extent (International Energy Agency 1983b).

Environmental disturbance The surface structures of an underground coal mine initially produce no more environmental disturbance than most large industrial plants. With the passage of time, however, the removal of material from underground may lead to surface subsidence with attendant effects on the built environment and landscape appearance.

The effects of surface mining, on the other hand, are immediate and extensive. It destroys the existing vegetation, may bury or mix the soil layers with parent material and significantly alter the topography (and thus the drainage system) over relatively large areas. In most countries environmental regulations require that the worst of these effects be minimized by selective overburden removal practices and reclamation procedures.

Residuals At the primary production stage, residuals associated with underground mining include acidic mine-water, mine-spoil heaps (and associated leachates) and dust. Overburden is the major residual from surface mining and particularly in hilly areas, overburden dumps and their run-off water can result in serious environmental degradation. Residuals associated with coal preparation plants include dust, turbid and acidic water and, with later processing stages, gases, liquids and particles. During the transportation and storage phases, residuals are generally limited to dust and 'noise'.

In new facilities all these residuals can be reduced to generally acceptable levels by process adjustments, recycling or containment. The remaining productive life and efficiency of old facilities, however, are often insufficient to justify the capital or operating expenditures required to modify them to acceptable standards.

Organizational structure

The organizational structure of the coal supply system has little of the vertical integration of the oil industry. In most countries, the resource base is owned by the state and the rights to explore and mine are obtained through leases.

However, in some parts of Western Europe and the USA coal ownership was acquired by the private sector in the early days of the evolution of the industry (Nelson 1983). In western Europe many of these rights have been extinguished in favour of the state. In North America and, to some extent, Australia, leases to the coal resource have increasingly been acquired by petroleum and metal mining companies, electric utilities and iron and steel companies. In the USA coal subsidiaries of petroleum companies are said to control approximately one fifth of recoverable reserves and one third of the production (Chapman 1983). In western Canada, Shell and British Petroleum have large holdings along with major metal mining companies such as Dennison and Teck corporations.

Mines and preparation plants in many countries (UK, France, India and the Centrally Planned Economies) are owned by nationally controlled coal agencies whereas in North America, Australia and South Africa they are privately owned. The links of the transportation system are rarely owned by the same agencies or companies as the production facilities although there is often state ownership of railroads and port facilities.

Petroleum

Petroleum refers to all hydrocarbons which occur in nature whether in gaseous, liquid or solid form. The phase in which they exist varies in response to temperature and pressure. Consequently, they may be in one phase underground and another at the surface and, during processing, they can be separated into fractions by manipulating the two variables. For convenience, only the liquid and solid forms will be dealt with here leaving natural gas to the next section.

Functional components

The production chain shown in Figure 4.2 illustrates the functional components of the conventional oil industry with the structure of the still little developed oil sand and shale industries added for comparison. The additional complexity (and thus cost) at the primary production end of oil sand and shale is noteworthy.

Resource base

Petroleum consists mainly of hydrocarbon compounds with small amounts of other chemicals and metals. The techniques and rate of production, the transportability and the refining processes and yield of products are dependent upon the physical and chemical properties of each crude oil source. These properties vary widely between individual sources with the result that some crude oils are much more valuable than others.

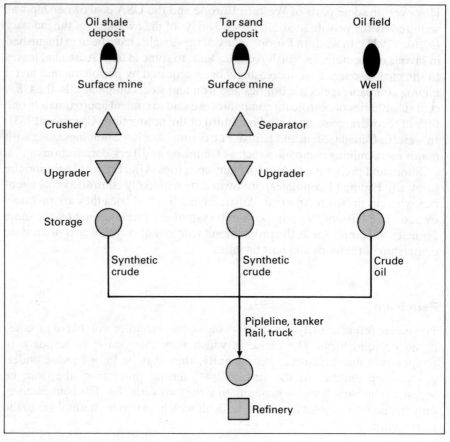

Figure 4.2 Oil supply chains

Characteristics and classification The most basic physical property of crude oil is its specific gravity which influences both its viscosity and heating value. Table 4.4 illustrates the range of specific gravity expressed in terms of the American Petroleum Institute (API) scale, a low value of which indicates a high specific gravity.

The viscosity of oil plays an important part in determining its flow characteristics which, in turn, affect production rates and ease of movement in pipelines.

The chemical properties which largely determine the relative values of crude oils include (a) the impurities present, (b) the carbon/hydrogen ratio and (c) the atomic and molecular structure. The most common impurities are sulphur, nitrogen and compounds of vanadium, nickel and iron. Low sulphur crudes (less than 2 per cent sulphur) command a premium price because of increasingly severe and widespread sulphur emission standards at the point of combustion.

Table 4.4 Specific gravity and heating value of petroleum and petroleum
products

Specific gravity API scale	Approx. heating BTU/lb	Commodity
10	18,500	11 API Bitumen
		10–15 API Heavy crude oil
		15 API Heavy fuel oil
20	19,000	
30	19,400	30–40 API Light crude oil
		35 API Average for US crude oil
40	19,800	40–45 API Light fuel oil
50	20,000	
60	20,300	65 API Gasoline
70	20,500	

Metallic impurities in crude oils impose constraints on the processes which can
be used to refine them.

There are over 5000 hydrocarbon compounds which differ from each other
in terms of carbon/hydrogen ratio and atomic–molecular structure. Crude oil
may also be classified in terms of the conditions of occurrence into *conventional*,
heavy, *tar sand* and *shale oil*. The first refers to crude that has a low viscosity
which permits it to flow to the surface whereas heavy oil, though still liquid, is
so viscous that it will not flow without special treatment. Tar sand oil is the
heavy asphalt (bitumen) which occurs as a plastic solid mixed with sand, and
shale oil refers to the solid organic material called kerogen which decomposes
into oil when heated.

Exploration and delineation   The discovery and assessment of conventional
oil involves an elaborate sequence of exploration, delineation and evaluation
procedures. Exploration may be divided into two categories: pre-drilling and
drilling. The former consists of searching for suitable geologic regions by means
of surface geological mapping followed by seeking favourable underground
geological structures within those regions by means of geophysical techniques
(seismic, gravity and magnetic). When the inferential evidence is sufficiently
encouraging, 'wild cat' drilling is employed to determine whether oil is actually
present. The drilling phase of exploration is much more costly than earlier
stages (Fuller 1984). There has been a steady development of the technology of
exploration providing an increasing variety and accuracy of information.
Furthermore, since the 1950s the use of the techniques has been extended to
offshore areas in ever increasing depths of water as well as into harsh Arctic
environments.

Once a discovery has been made the vertical and horizontal extent of the
deposit has to be determined, the potential production mechanisms and rates
estimated and the quality of the oil analysed. Only if these procedures yield
positive results will a production program be considered.

Production

The production of conventional and heavy oil by means of wells differs markedly from the technology used in the tar sand and oil shale industries. In the latter very large open pit mining and elaborate beneficiation replace the often inconspicuous oil well.

Conventional and heavy oil   Almost all the current world production of crude oil comes from metal tube oil wells inside a metal casing penetrating into the oil-bearing sediments. The oil flows up the tube at a rate determined by the viscosity of the oil, the permeability of the strata in which it occurs and the pressure at depth. Production rates may vary from only a few barrels per day to many thousands.

Under natural conditions, on average, only one-third of the oil in place in the reservoir is produced. In order to improve this *primary* recovery rate and thus prolong the productive life of the field as well as increase the current reserves, it is increasingly common to make use of enhanced recovery techniques. *Secondary* recovery usually refers to pumping and pressure maintenance procedures in which water or gas is pumped into the oil-bearing strata in order to force more oil to the surface. *Tertiary* recovery applies to processes designed to decrease the viscosity of *in situ* oil and to reduce the capillary forces in the oil-bearing rock. The most common procedures for decreasing the viscosity involve the injection of large quantities of high pressure and temperature steam. Mixing fluids and solvents also may be pumped into the oil-bearing formation to decrease capillary forces. These enhanced recovery techniques themselves use a significant amount of energy which, along with the capital investment required, make the oil produced more expensive than that produced by primary methods.

Tar sand oil   Production of bitumen from tar sands is currently only carried on in Canada although similar deposits occur elsewhere, particularly in the Orinoco Basin of Venezuela. Tar sand is mined by open-pit methods with the bitumen separated from the sand and upgraded by hydrogenation into a good quality synthetic crude oil. In the Canadian operations approximately 2.5 tonnes of sand plus varying amounts of overburden have to be removed to yield one barrel of synthetic crude. Commercial scale technology to produce bitumen from deposits at depths beyond the limits of open pit mining is not yet available.

Shale oil   Except for some production in the USSR, little commercial use has been made of the oil shale deposits of the world. Intensive research and pilot plant development have been undertaken in Wyoming and the adjacent states in the USA but no commercial scale plants have been built. The production chain involves large-scale underground mining (approximately 4 tonnes of shale per barrel of oil) followed by crushing and retorting to produce synthetic oil and, as with tar sand processing, very large amounts of sand residue.

Processing

Crude oil, whether natural or synthetic, is of little use until refined into petroleum products. The oil refining industry is one of the largest (in terms of capital investment), most widespread and technologically complex industries in the world. The yield of products from a refinery (Table 4.5) is a function of (1) the character of the crude oil input, (2) the equipment installed and thus the processes available and (3) market requirements. A simple refinery with, for example, only atmospheric distillation equipment and using one quality of crude oil will be able to produce only a limited number of products and have little or no flexibility in determining the amount of each. At the other extreme, a complex, multi-process refinery will be able to produce a complete range of products and vary (to a considerable extent) the proportion of each.

Table 4.5 Processes and products of the oil refining industry

		Processes	
Preparation	*Separation (distillation)*	*Compound splitting (cracking)*	*Compound building (reforming)*
Desalting	Atmospheric	Hydro-cracking	Alkylation
Dehydration	Vacuum	Thermal	Isomerization
		Catalytic	Polymerization

	Products		
Fuels ⎡ Butane ⎤ LPG ⎢ Propane ⎦ ⎢ Naptha ⎢ Gasoline ⎢ Kerosene ⎢ Diesel oil ⎣ Fuel oils Lubricants Solvents	Asphalt Petroleum coke Waxes	Petro-chemical feedstocks	⎡ Ethylene ⎢ Propylene ⎣ Butylene

Refineries range in capacity from as little as 1000 barrels of crude throughput per day up to almost 600,000 barrels. Although the refining process requires considerable amounts of heat, this is usually entirely supplied from the oil being processed. Nevertheless, the amount of the product stream used for fuel is only a small proportion of the total volume flowing through the plant (approximately 5 per cent). As a result, oil refining does not result in a large weight loss which is significant when considering the location of the industry.

Transportation and storage

The transportation system for oil and oil products consists of three major stages. The first, *gathering*, refers to the movement of crude from each well in a field by pipeline to a central point either for refining or, more usually, storage at a bulk shipping terminal. *Bulk shipment*, the second stage, is the large volume, long distance transfer of crude and products by tanker vessels and pipelines while the third stage, *distribution*, is the movement of products from refineries to intermediate or end users. This latter stage is carried out not only by vessels and pipelines but also by trains and trucks.

For overland transportation the pipeline is the most efficient mode of bulk shipment with a throughput determined by the viscosity of the oil, the diameter of the pipe and the pressure used. Diameters range from 15–120 cm (6–48 in) with daily throughputs of 35,000–1,000,000 barrels per day and large trunk lines extend for distances in excess of 3200 km (2000 miles) in North America and the USSR. Major lines cost $300,000 to $1,000,000 per kilometre ($500,000–1,500,000 per mile) to build in 1985 making it necessary to operate them close to capacity for relatively long periods if the investment is to be recovered and an acceptable rate of return achieved (McCaslin 1986, p. 343). Being fixed, linear facilities pipelines do not have the same spatial flexibility of the other major bulk carrier, the ocean-going tanker. Developments in pipeline technology include higher pressure operation and reversible flow capabilities and the ability to transport mixed batches of liquid fuels.

Although pipelines are used for the international transport of petroleum (e.g. Canada to the USA, USSR to Eastern Europe), the bulk of international movement is by tanker. Since 1950 the world tanker fleet has expanded from approximately 25 million DWT to over 300 million and the average size of vessel has increased from 12,000 DWT to over 100,000 DWT. The very largest crude carriers exceed 500,000 DWT and carry approximately 4 million barrels. Subject to terminal facilities with water deep enough to accommodate them when loaded, tankers are a spatially flexible mode of transport. The upward trend in tanker size appears to have levelled off and the next generation is expected to consist of vessels of 200–300,000 DWT. Other developments are likely to be toward more efficient propulsion systems, greater tank segregation and heating coils to improve flow characteristics of heavy oils during loading and unloading (Champness 1981).

Storage of crude oil and products is needed to meet peak demands, to protect transporters, processors and users from supply interruptions and as a protection from upward price changes. Thus storage facilities of varying capacity are located at the point of production, loading and unloading terminals, refineries, as well as at wholesaling, retailing and end use locations. Steel tanks are used for large-scale storage either above or below ground, giving rise to the familiar 'tank farm'. Individual tanks may have a capacity in excess of a million barrels. In addition, oil may be stored in fleets of non-operating tankers and in natural underground structures such as worked-out salt domes, natural caverns and disused mines. The majority of the Strategic Petroleum Reserve of the USA is in such structures.

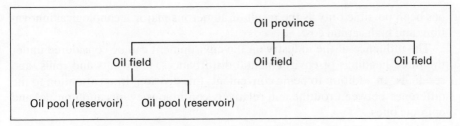

Figure 4.3 Hierarchy of terminology for oil resource occurrences

Spatial characteristics

The hierarchy of terms used to describe the spatial distribution of conventional oil resources is shown in Figure 4.3. An *oil province* refers to a region in which the geological conditions are generally favourable for the occurrence of oil (e.g. the Persian Gulf) or sometimes, and more specifically, to a part of a larger region such as the Salt Dome Province of the US Gulf Coast (Parker, S. P. 1981). The term *oil field* is used to describe a group of pools or reservoirs which may be distributed horizontally or vertically. Examples include such fields as Ghawar (Saudi Arabia), Forties (North Sea) and East Texas (USA). An *oil pool* or *reservoir* refers to a particular oil-bearing sedimentary formation rather than a concave container full of oil as the term suggests.

Oil is spatially associated with porous rocks such as sandstones and lime-stones originally laid down in water rich with organic life. Subsequent pressure and temperature changes cause oil to form while later depositional or structural changes form traps which prevent the oil from dispersing.

Because of their high energy content and liquid form (though with variable viscosity) crude oil and its products are readily transportable by a variety of transportation modes. These characteristics, together with the very large variety of uses and widespread dispersion of users, result in extremely large-scale and complex spatial interaction systems. Crude oil and its products constitute the largest item (by weight, volume and value) in international trade and, in most developed countries, account for a major proportion of the tonne-kilometres generated on internal transport systems. Indeed at all scales, petroleum transport networks are among the most fully developed and technologically advanced circulation systems in use today.

Environmental aspects

Environmental aspects of the oil supply system include consideration of both the influence of the environment on the industry and vice versa. The environmental conditions in which oil exploration, production and transportation are carried on have significant technical and cost implications. As the industry has extended into offshore and Arctic environments, many constraints have had to be overcome. The rapid increase in activity in these two types of environment has been possible only as the result of numerous major technological innovations and high capital cost.

The influence of the industry on the environment can be considered under the three headings of environmental disturbances, blow-outs and spills, and residuals. In addition to being convenient, this division draws attention to the difference between routine and relatively continuous events and unusual and episodic ones.

Environmental disturbance   During exploration, environmental disturbance is relatively localized and limited to scarring by heavy, tracked vehicles and disturbance of vegetation and drainage. Some production facilities can be surprisingly unobtrusive and blend into urban and rural landscapes; others, particularly in the past, dominate the scene with a forest of derricks and other production equipment. In some areas, post-production subsidence results in significant environmental disturbance.

In the construction of pipelines, a narrow swath of land is affected but restoration usually returns the surface to its original state except in forested areas where the right-of-way remains conspicuous. Laying pipe across rivers requires particular care to avoid large-scale disturbances, the effects of which may be carried long distances downstream. A few pipelines are constructed above ground (e.g. Aleyska in Alaska) and may constitute a barrier to the migration of wildlife.

Blow-outs and Spills   During the drilling of exploration or production wells unexpectedly high pressures may be encountered causing a blow-out during which oil flows uncontrolled to the surface and escapes into the environment until the well can be capped. On land the oil can usually be contained in a relatively small area and the flow brought under control quickly. Undersea blow-outs are a much greater hazard because the oil cannot be contained and the difficulty of capping a well on the bottom of the ocean may mean that the uncontrolled flow continues for a long time. A blow-out in the Bay of Campeche, Mexico, remained uncontrolled for almost 10 months in 1979–80, spilling approximately 3 million barrels of crude oil (*International Petroleum Encyclopedia* 1981).

Oil spills are the result of a rupture of the container in which the oil is being transported. Large scale incidents on land (e.g. the sabotage of a pipeline) can be contained and rapidly controlled, generally limiting their effect to local areas. At sea, collisions or groundings of tankers can lead to large quantities of oil spilling onto the water and spreading in response to wind and water conditions. In 1978, the *Amoco Cadiz* ran aground on the north coast of Brittany, spilling 1.6 million barrels of crude oil (Carter 1978).

When large-scale blow-outs and spills occur they are dramatic and well publicized events which, not surprisingly, lead to considerable public apprehension. More surprising, considering the extraordinarily large-scale and widespread distribution of oil transportation, is that so few such events occur.

Residuals Few residuals arise from exploration and delineation activity beyond camp wastes and abandoned equipment but, during production, the spent drilling muds and entrained brine may pose disposal problems in environmentally sensitive areas. Tanker transportation leads to routine tank cleaning and the unloading of ballast water which is one of the major sources of hydrocarbons in the oceans (Fowler 1984). Noxious and potentially harmful gaseous and liquid residuals are associated with refining as the result of the removal of impurities and the combustion of fuel to provide the requisite process heat. However, in modern state-of-the-art refineries the majority of these wastes are contained in-plant and treated to avoid environmental degradation.

Organizational structure

The organization of the oil industry has changed dramatically since approximately 1970 (Hartshorn 1980). In order to summarize these changes and briefly outline the current scene, it is useful to consider the structure in terms of three dimensions: the type of companies involved, their vertical integration, and the spatial scale of their operations. Companies may be divided into two major types, each with its own subdivision. First, there are private companies, the shares of which trade on the major stock exchanges of the world and which may be subdivided into multi-nationals and 'independents'. The former have a considerable degree of vertical integration and operate on a worldwide scale, while the latter are usually only partially integrated and regional or national in scope.

Secondly, there are companies owned by the public sector, 'national' companies. Governments of countries which are both net importers and net exporters of oil have formed oil companies as one means of achieving national energy objectives. Such companies have varying degrees of vertical integration and, by definition, are initially national in scope though a number of them have expanded their operations beyond their own borders.

In 1960 the world oil industry (excluding the USSR) was dominated by seven multi-national companies (Exxon, Mobil, Texaco, Chevron, Gulf, Royal Dutch Shell and British Petroleum), five of which had their headquarters in the USA and two, Shell and British Petroleum, in the Netherlands and United Kingdom respectively. These companies exercised almost total control of all phases of the oil industry and constituted the largest commercial organizations in the world. They managed virtually all international trade in oil and controlled (by means of concessions from the governments of the countries with the resources) the majority of the non-communist oil resources (McCaslin 1982). Furthermore, their oil-producing capacity exceeded their refining capacity leaving them surplus crude which they sold to independents and the national companies of consuming countries (Mohnfeld 1984).

By the mid-1980s the organizational structure had changed dramatically.

National companies of the producer countries (particularly members of the Organization of Petroleum Exporting Countries, OPEC) had emerged as the dominant organizations in the production phase of the industry and some were also rapidly moving into downstream operations. As a consequence, the private multi-national companies no longer control the world industry. Even during a period of general reduction of refining capacity such as in the mid-1980s, many of them do not control enough crude oil production to meet their refining needs and must buy from the national companies and the spot market. International trade in oil and oil products has become somewhat less flexible as bilateral agreements between the national companies of producer and consumer countries replace the complex commercial arrangements of previous decades (Mohnfeld 1984, p. 168).

Natural gas

Natural gas is a naturally occurring mixture of combustible hydrocarbons and other gases. Suitable geologic environments for the occurrence of natural gas occur in every system of rocks down to the Cambrian although conditions leading to large concentrations of producible gas are more restricted.

Functional components

As natural gas and crude oil are the gaseous and liquid phases of the same family of compounds (hydrocarbons) it is to be expected that the two industries would be closely linked. Indeed it is common to use the term 'petroleum industry' to refer jointly to the oil and gas industries. Despite these inter-relations, the natural gas industry is sufficiently distinct to warrant treatment in a separate section in this chapter although, to avoid duplication, some parts of the complete production chain (Figure 4.4) are omitted (e.g. exploration and delineation).

Resource base

The natural gas resource base is commonly described in terms of conventional and unconventional sources. *Conventional* sources refer to concentrations of gas that occur in sedimentary formations to depths down to approximately 8000 m (25,000 ft) and which have porosity and permeability characteristics that permit the gas to flow to the surface at a commercial rate. Sources such as these are further classified into associated and non-associated categories. The former refers to gas which occurs in *association* with oil, either dissolved in the oil ('casing head gas') or in contact with the oil but concentrated in a 'gas cap' above the oil-bearing strata. *Non-associated* gas, as the term implies, occurs independently of oil.

The term *unconventional* is used to describe gas that occurs under cir-

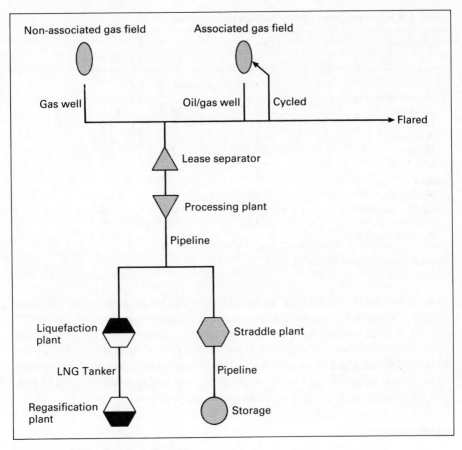

Figure 4.4 Natural gas supply chain

cumstances which are not exploitable under current technical or economic conditions. Five such unconventional sources are commonly identified: (1) tight sands with permeability too low to permit commercial production without 'enhanced' recovery techniques, (2) coal-bed gas, (3) gas shales, (4) gas dissolved in subterranean water (geopressurized gas) and, (5) at low sub-surface temperatures, natural gas hydrates in which gas is trapped in ice (Mankin 1983; Holder 1984).

Characteristics and classification Natural gas consists mainly of methane, other hydrocarbons and such other gases as nitrogen, carbon dioxide, helium and hydrogen sulphide (Table 4.6). The proportion in which these constituents are present varies considerably from one field to another. It is common to distinguish between dry and wet gas and sweet and sour.

Dry and wet gases differ from one another on the basis of the amounts of hydrocarbons other than methane that are present. Dry gas is almost entirely methane whereas wet gas contains significant proportions of higher hydrocar-

85

Table 4.6 Representative analyses of raw natural gases

Constituents	Dry	Wet	
		Sweet	Sour
		Percentage	
Hydrocarbons			
Methane	95.6	84.7	63.6
Ethane	0.2	6.1	4.0
Propane	0.1	2.3	2.5
Butane	0.1	2.4	2.1
Other	—	—	6.7
Other			
Nitrogen	3.7	4.0	1.5
Carbon dioxide	—	0.5	4.2
Helium	0.3	—	—
Hydrogen sulphide	—	—	15.4

Source: Simpson and Rutledge 1964, 25

bons collectively referred to as natural gas liquids (e.g. ethane, propane, butane). These latter compounds are in gaseous form underground but readily liquefy at near surface temperature and pressure conditions. They have high value as fuel (butane has a heating value 3 times that of methane) and as a basis for petrochemicals. The second distinction is made on the basis of the amount of hydrogen sulphide (and, thus, sulphur) present and results in *sweet* and *sour* gas. The latter, high in sulphur, is corrosive and noxious.

Production

Natural gas is produced in essentially the same manner as conventional crude oil by means of wells drilled into the productive strata. Indeed, as has already been noted, associated gas is unavoidably produced in conjunction with oil and thus is produced at a rate determined by the oil-lifting schedule. The production of non-associated gas, on the other hand, is managed independently and if there is no market the well is capped and the gas becomes 'shut in'.

Associated gas is either flared, recycled back into the petroleum-bearing strata or marketed. For decades some of the world's great oilfields were visible from great distances at night on account of the flaming gas flares associated with each oil well. Untold quantities of this valuable fuel have been (and still are in some areas) dissipated by this wasteful practice. Recycling the gas back into the ground helps to maintain the pressure in the reservoir for driving the production of oil and preserving the natural gas liquids (Harder 1982).

Processing

Before being marketed, all gas containing natural gas liquids or requiring purification is treated in one or more processing plants. The first stage takes place in a *lease separator* where the constituents that condense at atmospheric temperature and pressure are removed. More elaborate processing occurs at the second stage in a *gas processing plant* which removes most of the natural gas liquids (NGL) and, where present, hydrogen sulphide and other undesirable constituents. Propane and butane are marketed as fuels (in pressurized metal containers) or feed-stocks for petrochemical purposes. Other liquids include natural gasoline, pentanes-plus and ethane. Not all the NGL have to be removed before the gas enters the pipeline for distribution. In fact ethane and small amounts of propane and butane (all of which have higher heating value than methane) are deliberately retained in the gas in order to maintain its heating value at a specified level.

Two other types of gas-processing plants should also be mentioned. Some countries (e.g. Canada) have natural gas which is particularly rich in ethane, the resource base of ethylene, a valuable petrochemical feedstock. In such cases, facilities known as *straddle plants* process very large quantities of pipeline gas solely in order to remove ethane. The second class of special processing facilities includes those required for the liquefaction of gas referred to in the next section.

Transportation and storage

The overland transport of natural gas is almost exclusively carried on by pipelines whereas most NGL move in pressurized tanks. The capacity of a gas line is a function of the diameter of the line and the pressure at which it is operated which, in turn, is a function of the structural characteristics of the line and the size and number of compressor stations. Gas pipelines range in diameter from 2.5 cm (1 in) domestic distribution lines all the way to the giant 142 cm (56 in) trunk transport line from Siberia to Western Europe.

The technology of overland pipelining of gas is highly developed and elaborate systems extend over North America, Western Europe and the USSR. Oceanic transport, however, has been slower to develop. Techniques have been introduced for relatively short underwater pipelines from offshore fields and even across the Mediterranean but long-haul oceanic transport had to await the development of liquefied natural gas (LNG) technology. When liquefied at a temperature of about $-127\,°C$ ($-260\,°F$), natural gas occupies only 1/600 of its volume at atmospheric pressure thus raising the energy content per unit volume approximately 600 times. In this form it is economically feasible, but still expensive, to transport gas in cryogenic insulated tanks. LNG transport systems require a liquefaction plant at the loading terminal, LNG tankers of 125,000–150,000 m^3 capacity and a vaporizing plant at the unloading terminal.

Liquefaction may also be used to store natural gas and cryogenic tanks are commonly used by gas distribution firms for this purpose. Another commonly used storage technique is to pump gas underground into impermeable geologic structures.

Spatial characteristics

The chief difference between the spatial characteristics of oil and natural gas arises from the much greater costs of moving gas inter-continentally. Oil and oil products are transported extensively at this scale but, despite the availability of LNG technology, the amount of trans-oceanic movement of gas in the mid-1980s was not as great as was earlier expected. Overland, however, large diameter long distance pipelines have permitted considerable intra-continental movement (e.g. Siberia to Western Europe and within North America). At the urban scale, complex networks of small diameter distribution lines supply individual users on a continuous basis without the need for user storage facilities.

In terms of spatial association there is another difference between oil and gas. As noted in the preceding section, most natural gas has to be processed (at least to some extent) *before* it enters a long distance transportation system whereas crude oil may move long distances before it is refined. Gas processing yields valuable raw materials for the petrochemical industry. Consequently, gas resource areas may be associated with advanced stages of industrial development to a greater degree than many oil resource areas which simply produce oil and ship it in crude form.

Environmental aspects

The natural gas supply industry is generally the most environmentally benign of the fossil fuel industries. Although the exploration and pipeline construction phases have the same impacts as in the oil industry, gas processing plants produce fewer noxious wastes than oil refineries (though the removal of sulphur can lead to hazards) and gas transportation accidents have only local, even if sometimes dramatic, effects.

Organizational structure

Despite the close association between the oil and gas industry at the production level, the two industries display some notable organizational differences. First, there are many more companies in the gas business and, secondly, the firms are characterized by much less vertical integration. The major oil companies (both private and national) are often large gas producers and primary processors but they are much less frequently involved in the transportation or end use marketing phases. In North America many gas companies have their capital origins in diverse, non-energy enterprises (e.g. Dome Petroleum in Canada) or have entered the business as pipeline companies and moved upstream to secure a resource base.

Uranium

Uranium is the most recent addition to the list of primary energy sources from which humans have learned to recover useful energy. Nuclear fuels manufactured from uranium are unique in two respects: they have an energy/weight ratio which is orders of magnitude greater than any other fuel and they, along with the equipment used in processing and converting them, are potentially lethal to all forms of life for extended periods of time.

Functional components

The uranium supply chain is not only technologically more complex than other supply chains but it is also different in kind. Combustion residuals from fossil fuels are noxious and if dumped into environmental systems in sufficient quantities are harmful but only rarely are they lethal. Residuals from nuclear fission are lethal and cannot be allowed to enter the environment directly. Consequently it is more appropriate in this section to consider the full nuclear fuel cycle (Fig. 4.5) rather than just the supply chain up to the point of end use. This serves to emphasize the commitment that has to be made to permanent storage and/or reprocessing of nuclear residuals.

Resource base

Uranium is found in a variety of minerals and geological conditions in the earth's crust and in ocean water. Although the geological concentrations are greater than oceanic, both are very low. Land deposits containing less than

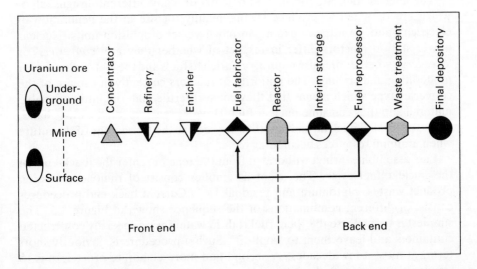

Figure 4.5 Nuclear fuel cycle

1 per cent of uranium oxide are considered good grade and some with less than 0.1 per cent are mined.

The element uranium consists of three isotopes (species of atoms which differ from one another in terms of their mass and thus their physical behaviour), U^{238} (99.28 per cent), U^{235} (0.71 per cent) and U^{234} (0.01 per cent). Only U^{235} is fissionable and thus suitable for reactor fuel. Less than 1 per cent of the ore mined is uranium and, of that amount, less than 1 per cent is U^{235}.

Production and processing

In what is termed the front end of the nuclear fuel cycle, uranium ore is produced by both underground and surface mining techniques and then undergoes three or four stages of processing before being available for use in a reactor:

Concentration: at the mine site the ore is milled to produce a concentrate of uranium known as 'yellowcake'.

Refining: the concentrate has to have impurities such as boron and cadmium removed at a refinery.

Enrichment: the product of the refinery still has the natural proportion of the three major isotopes, only one of which, U^{235}, is fissionable. The enrichment process increases the proportion of this isotope and thus improves the quality of the nuclear fuel. Although most reactors require enriched fuels, one, the CANDU, does not.

Fuel fabrication: the final stage of processing is the manufacture of fuel elements, made up of pellets of UO_2 contained in tubes of zinc alloy or stainless steel and assembled into bundles.

The nuclear fuels are 'burned' in reactors of many different designs, all of which are designed to produce steam, mainly for use in the generation of electricity and, to a lesser extent, in naval vessel propulsion units. Nuclear power station reactors differ in respect of whether they are '*converters*' or '*breeders*', whether they use natural or enriched fuels and the materials used for moderating and cooling. The majority of reactors currently in use are of the converter type which means that they convert fertile material into fissionable material and thus consume nuclear fuel. Breeder reactors, on the other hand, create or 'breed' nuclear fuel and thus are seen as the technology of the future when uranium resources become scarce.

Fuel assemblies, when withdrawn from the reactor, enter the 'back-end' of the nuclear fuel cycle. The spent assemblies consist of radioactive fission product wastes, plutonium and residual U^{235}. Current back-end procedures consist of different combinations of the sequence shown in Figure 4.5. The simplest is to immerse the spent fuel rods in water inside specially constructed containers and leave them to 'cool-off'. Such a procedure is obviously short term, at most a few decades. A second and more elaborate procedure is to remove the rods from the interim short-term storage and reprocess them to remove the reusable U^{235} (for recycling to fuel fabrication plants) and plutonium. The residual liquids may then be solidified and embodied in glass in waste treatment facilities and once more put into either interim or final storage. Fuel reprocessing plants are largely operated by remote control, are very capital- and energy-intensive, and are among the largest industrial plants built and yet actually process only a few tonnes of material per day.

Transportation and storage

The transportation of nuclear fuel components differs from that of the fossil fuels in two important respects. First, the amounts transported are measured in hundreds of kilograms not thousands of tonnes, millions of barrels or billions of cubic metres. Secondly, most of them have to be completely isolated from the environment and carried in special containers which provide total containment under normal operating circumstances and are immune to rupture in the event of any conceivable accident. These two characteristics lead to transport by truck and rail although the two products at either end of the complete cycle (yellowcake and vitrified wastes) are also transportable by ships. It is noteworthy that toxic nuclear materials have been moved millions of kilometres without one recorded accident involving the release of radioactivity.

The storage requirements of the nuclear supply chain are also different from those of fossil fuel chains. Except at the mine-located concentrators where, because of the low grade of uranium ores, the amount of tailings is large, the volume of material to be stored is very small in comparison with fossil fuels. However, because of its extremely toxic nature the effort that has to be expended to confine it and isolate it from the biosphere is considerable and costly.

Spatial characteristics

If there is one word which typifies the essential character of the spatial distribution of the nuclear fuel cycle it is 'concentration'. Although the mining stage is most widely distributed, there are still only approximately a dozen countries in the world producing uranium and fewer than twenty-five with installed nuclear reactors. Refining, enrichment and fuel fabrication plants are limited to less than ten countries and reprocessing facilities to no more than five or six.

Because of the hazardous nature of the materials involved some countries and states do not permit nuclear facilities of any kind (e.g. New Zealand and the province of British Columbia in Canada). Such restrictions are becoming more common in local communities and the specific siting of nuclear reactors, processing facilities and storage areas not only have to meet exacting technical specifications but increasingly selective community values.

In terms of spatial interaction, the transport of uranium products is greatly restricted in volume and extent in comparison with the fossil fuels. Bulk transportation systems of the sort used by the coal or oil industry are just not required.

Environmental aspects

In all stages of the nuclear fuel cycle (except mining and milling) the material being handled, the equipment used and the residuals must be isolated from the biosphere. Even during mining and milling special care must be taken to protect miners from radon gas and to contain tailings and associated leachates. Nuclear materials and residuals may be broadly distinguished on the basis of whether they emit low or high levels of radiation and whether the level persists for short (months, years), intermediate (decades, centuries) or long (10^3–10^6) years) periods. The intensity of radiation determines the degree of containment required and the persistence determines the length of time containment and isolation from the environment must be provided.

At present, nuclear residuals are contained and placed into short-intermediate water-filled stainless steel and concrete tanks usually at the site of the plant at which they were produced. Long term or permanent storage sites are not yet in use. Investigations are under way to determine the suitability of storage either in underground cavities in impermeable, completely stable geological structures or burial in red-clays at the bottom of ocean deeps. Before the end of this century some of the early nuclear reactors used in the generation of electricity will be de-commissioned raising the question of their disposal. Temporary measures involve shutting down the plant and the establishment of a complete security system to prevent re-entry. Long term measures require partial or complete dismantling and the entombment of equipment in a massive cocoon of reinforced concrete.

Organizational structure

The mining and milling of uranium ores, outside the Centrally Planned Economies, is in the hands of large private mining companies, some of them divisions of energy conglomerates. The remaining stages of the front-end of the nuclear cycle are still generally carried out by government controlled agencies or corporations. The reactors themselves are owned by electric utility companies and the back-end of the cycle (to the extent that it is developed) involves a mix of private and public sector organizations. Research and development is also shared between the two sectors (with the emphasis on the public sector) whereas the design and construction of reactors and ancillary equipment is largely carried out by a relatively small number of international-scale private companies (e.g. General Electric, Westinghouse).

Because of the hazardous nature of its operations and the potential use of nuclear materials for weapons, the nuclear fuel industry is the most widely and stringently regulated of the energy industries. International and national agencies attempt to control and monitor every stage of the nuclear cycle by means of continuously evolving regulatory and licensing procedures and, in the case of national governments, by direct participation.

Renewable sources

In addition to the non-renewable, mineral energy sources upon which the world depends so heavily, there are a number of renewable sources, some in current use, others awaiting technological developments. Renewable resources are inherently attractive because the spectre of exhaustion is not present, they have fewer environmental implications (or at least more benign) and provide an opportunity to decentralize the geography of energy supply. However, they also have qualities which pose development problems such as seasonal and diurnal variations, intermittency, generally low energy content (thus large land requirements) and very little potential for transfer in their natural state. In this concluding section on primary energy supply chains the major renewable sources are reviewed in a summary way. The topical divisions used in the previous sections are addressed but not individually identified by sub-headings.

Water power

The largest scale renewable energy source developments are associated with water power. While rivers are the major natural systems in which a mass of water has a large potential energy, popularly referred to as hydro-power, it is convenient also to include tidal systems in the general category of water power.

Hydro-power   A large waterfall is one of the most dramatic natural demonstrations of sheer power (and beauty). For many centuries humans have harnessed such potential energy and the less dramatically visible energy of swiftly moving streams by means of the water wheel but the large scale hydro-electric developments are products of the last 50 years.

The theoretical power capacity of a river is a function of the volume of water and the vertical distance through which it falls (head). The relationship may be expressed as:

$$P = 9.81\ QH$$

P = Power in kW
Q = Flow in m^3/sec
H = Head in metres

or

$$P = \frac{QH}{11.1}$$

P = Power in kW
Q = Flow in ft^3/sec
H = Head in feet

The data necessary to establish the flow characteristics of a river and to develop a knowledge of the processes involved in its regime require the operation of a network of hydro-meteorological stations for a number of years. The resulting record establishes the range between high and low flow both annually and

93

seasonally. In many rivers, high flow exceeds the low by many times, rendering them less suitable for development than those with a more balanced regime. Large high/low flow ratios may be reduced by creating reservoirs behind man-made dams. The head of water is a function of the river gradient; the steeper the gradient (e.g. waterfalls, rapids) the greater the natural head, the higher the dam the greater the artificial head.

Realization of the potential power capacity of a river to generate electricity requires the construction of facilities to capture the falling water and direct it onto a turbine. Conventionally this involves a dam or barrage, penstocks to transport the water and a power-house containing turbine-generator sets. High dams, backed by storage reservoirs, require firm and stable foundations and are usually built in gorges where the river bed is deeply confined. Low dams, or run-of-the-river facilities, which do not impound the water, are characteristic of large volume, low head rivers. The potential of a river may be developed at one site or a series of sites throughout its course. The largest single-site facilities commonly exceed 2000 MW capacity while complete river system developments may exceed 10,000 MW.

Environmentally, hydro-electric plants do not produce residuals in the process of generating electricity but the dam/power house structure usually results in considerable environmental disturbance, often in areas prized for their wilderness character and natural beauty. The major disturbances arise from the flooding of large areas upstream from the reservoir, the barrier effect of the dam itself and the alteration of the downstream flow in the river with potential implications for ecosystems and geomorphological processes. Such disturbances are not necessarily all negative as multi-purpose operating procedures may improve recreational potential and provide opportunities for flood-control and irrigation.

Tidal power A little-used form of water power is potentially available in coastal areas where large tidal ranges of 10 metres (30 ft) or more are amplified in relatively shallow inlets. The coastal topography must provide an area large enough to store the necessary volume of incoming water and yet include one or more narrow sections across which dams may be built.

Geothermal

In some geologically favourable areas of the earth's crust the normal rise of temperature with depth below the surface (approximately 30°C/km) is increased substantially, resulting in natural geothermal reservoirs of heat near the surface. Of the several types of reservoirs, the hydrothermal type containing hot water or steam (usually with considerable amounts of mineral impurities) is the only one in current use. Assessment of the resource requires knowledge of the temperature, depth of occurrence and water supply. Currently temperatures in the range of 50 °C are required for direct space heating use and over 200 °C for geothermal electric power stations.

The supply of geothermal heat is, indeed, continuously renewable but the supply of the water which transfers the heat to the surface when the reservoir is penetrated by a well is only continuous if it is currently being recharged from the surface. In some cases (e.g. Geysers Field, California) the water entered the reservoir in a previous geological period and is no longer being naturally recharged. Under such circumstances, the renewable heat source is effectively rendered non-renewable because the heat carrier (water or steam) is exhaustible.

The use of geothermal energy has only local environmental implications provided mineral-rich brines do not enter surface hydrological systems. The sight and sounds of production facilities degrade the environment especially if the area is one of natural beauty or a wildlife habitat. Furthermore, continued withdrawal of fluids may result in surface subsidence and disruption of production facilities.

Wind

The energy of a moving mass of air has been used for centuries to propel vessels and lift water and, in this century, to generate small quantities of electricity at dispersed locations. Today there is renewed interest in these capabilities. The power available from wind is a function of the wind speed and, for windmills (or wind turbines as they are increasingly called) the blade-tip circle diameter. Wind speed varies considerably from place to place on the earth's surface and at any one place varies seasonally, diurnally, hourly and over shorter time periods. Consequently, before any installations of a commercial scale can be considered, the wind regime must be assessed in terms of the persistency of various wind speeds. Even in the most windy locations the short-term variations in wind speed result in an uneven flow of primary energy and thus irregularity of delivered energy. To compensate for this, storage systems of various kinds must be provided, back-up energy systems must be available or consumers must adjust their use patterns to such irregularities. Because of the low density of air and the diffuse nature of wind energy the capacity of any one installation rarely exceeds 1000 kW.

Solar

Solar radiation is received at the surface of the earth either as direct or diffuse radiation. The amount of direct incoming radiation available at any given surface location is a function of latitude, season, time of day and clarity of atmosphere while the amount actually received on a specific surface varies in response to aspect, slope and intervening obstacles. Thus the supply of direct solar energy available to a user is at best intermittent and usually variable. Establishing the character of the solar regime of any site to support investment in equipment requires relatively long records of meteorological and radiative conditions. For much of the world such records do not exist either because

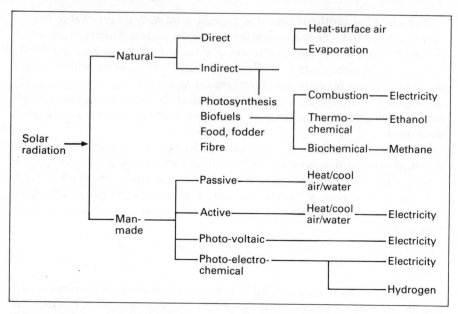

Figure 4.6 Solar energy supply chains

radiation measuring instruments have not been installed or the density of the network has been insufficient to yield data that can be used for site-specific design and operational purposes. The development of techniques to make use of data gathered at the mesoscale for application at the microscale is helping to reduce this constraint (Hay and Hanson 1985).

The many different routes by which solar energy currently and potentially contributes to the supply of energy for human use may be broadly grouped into natural and man-made systems (Fig. 4.6). The natural systems include direct contributions which we all experience to some extent and are manifested in air and surface heat and evaporation and the indirect input which, via photosynthesis, results in plant and other organic growth or biomass. Man-made systems include four identifiable routes: passive, active, photovoltaic and photoelectrochemical.

Natural solar energy systems – biofuels

In the twentieth century the major sources of commercial energy consist of organic material which grew for countless years in past geological time. This material decayed and became concentrated and changed in form to become fossilized biomass or the familiar fossil fuels. Today, as every season passes, the global ecosystems (terrestrial and aquatic) produce an increment of organic growth and contribute a supply of residuals to the decay organisms. These annual increments of biomass, the products of solar-driven photosynthesis, constitute valuable actual and potential sources of energy (biofuels) on a recurring basis albeit with very much lower energy content and extending over far greater areas than the fossil fuels.

The *productivity* of biofuel systems (e.g. forests and woodlands, grains and sugar cane for alcohol) in its simplest terms is a function of environmental conditions and the species and varieties of organisms concerned. Some environments provide conditions for continuous growth (e.g. humid tropics) while others are markedly seasonal. In the former, growth is rapid and continuous cropping schedules are possible whereas in the latter growth is slower and harvesting of planted products, periodic.

In the other energy supply systems considered in this chapter, the basic stock or flow of primary energy is pre-determined and beyond human control. The question is how much of the given stock or flow can be produced for human use? For biomass systems this is not the case. Even for natural (i.e. unmanaged) systems, human intervention determines their extent and, by influencing the ecological variables, the amount of material produced. In managed systems the kind and amount of biomass material produced is to a considerable extent determined by human decisions. Thus whether to use land to grow plants for fuel, fodder or fibre, what species and varieties to plant, how much to augment the natural supply of moisture and nutrients and to what extent to provide protection from pests and diseases are all matters of human choice which will greatly affect the production of biofuels.

Production of biofuels first requires the allocation of land to that purpose whether used by natural or managed systems. Secondly, production from natural systems involves harvesting methods ranging from the manual collection of debris and portions of standing plants to machine felling and harvesting of trees and crops. For managed systems, production is more complicated and embodies the whole chain of activities beginning with land preparation, through planting, tending and, finally, harvesting. Production of residuals which may be used for energy is, as the term suggests, a secondary matter following initial production for some non-energy purpose. Thus the straw remaining after the grain is harvested, the manure associated with a large scale cattle feed-lot operation, the garbage collected from a city or the sawdust and bark chips piled at a sawmill are the residuals remaining after some primary purpose has been served.

There are three pathways which may be followed to *process* biofuels to a state which renders them usable for energy: drying, thermo-chemical and biochemical. All biofuels initially contain some moisture which adds to their weight and diminishes their heating value. Once dried, either in the atmosphere or an oven or kiln, organic material may be burned in facilities as small and dispersed as a fireplace or cookstove or large and centralized as an industrial processing plant or electricity-generating station. The simplest thermo-chemical conversion is the processing of wood into charcoal which greatly reduces its weight and increases the heat/weight ratio thus decreasing transport costs (though increasing fragility). More elaborate thermo-chemical processing leads to the conversion of biofuels into gases and liquids while biochemical processing makes use of fermentation and anaerobic digestion to produce methane. Ethanol is being produced on a large scale in Brazil from sugar cane as a source of gasoline additive and the conversion of animal wastes into methane (biogas) is increasingly practised in China and India.

The most notable *spatial characteristics* of biofuel supply systems are the extensive areas required for growth and the limited range of distribution systems. These characteristics encourage small to moderate scale, decentralized processing and use facilities.

Environmentally, biofuel systems are often perceived to have few undesirable implications. This optimistic view overlooks such impacts as the deforestation associated with uncontrolled firewood gathering, the reduction of nutrient and organic content of the soil which results from the use of animal and crop wastes for fuel and the degradation of atmospheric quality associated with burning biofuels in domestic appliances. Furthermore, if biofuels were to be produced in large quantities, questions arise concerning the ability of the remaining land to produce biomass for other purposes, the ecological impact of widespread monocultural energy plantations and the implications of the potential demand for water and energy-dependent artificial fertilizers.

Organizationally, biofuel energy supply systems are quite different from fossil fuel systems. Extensive natural biosystems are frequently common property resources and, particularly in developing economies, allocation and enforcement of rights to harvest are either non-existent or ineffective. Biofuel supply systems in the Third World are often referred to as 'non-commercial' systems. However, a growing literature demonstrates that there is a complex web of small scale merchandising of both firewood and charcoal in such economies (Morgan *et al*. 1980).

For managed biofuel production the chief resource is productive land which, of course, may be owned or leased by individuals and public- and investor-owned corporations. To date few private corporations have entered the biofuel industry, although in some countries (e.g. Brazil, Sweden) both private and public organizations are moving into an organizational structure still largely dominated by individuals.

Man-made solar energy systems   Passive systems use built components to absorb, store and distribute heat. They may be used to heat (by absorption and convection) and cool (by convection, radiation and evaporation) water and air within low to moderate temperature ranges. Active systems are more complex and make use of separate collection, distribution and storage systems. In both systems the solar collecting surface may consist of a flat, non-focusing plate with operating temperatures of 90–150 °C or focusing collectors with operating temperatures which may exceed 1000 °C. The latter systems, with the added feature of equipment to track the sun, make more efficient use of the incoming solar radiation and show considerable promise as the basic technology for solar-thermal central electric stations.

Photo-voltaic and electrochemical technologies are two means by which the light component of the solar radiation spectrum may be converted into electricity. Photo-voltaic cells, usually consisting of silicon wafers, convert sunlight into direct current electricity. Photo-electrochemical cells consist of semi-conductors immersed in a liquid electrolyte which convert solar radiation into either electricity or hydrogen. Currently both solar-photo and solar-thermal means of generating electricity are more costly than conventional systems but may provide valuable back-up capacity in isolated locations or in special environmental circumstances (e.g. in space).

Electricity

The energy supply systems considered in the preceding sections of this chapter have all been based upon primary energy sources. However, an ever increasing proportion of fossil fuels, almost all uranium and varying proportions of renewable energy sources are converted into one secondary form of energy, electricity. Because of this trend and the fundamental importance and wide-spread use of electricity it is treated as a separate system in this final section.

Functional components

The commercial production of electricity involves the transformation of mechanical into electrical energy by means of a generator. In most electric supply systems the generators produce alternating-current (ac) electricity (rather than direct-current) at relatively low voltage. Although the convenience and versatility of electricity distinguish it from other forms of energy, it differs from them in other ways which significantly affect the structure of the industry: it must be used the instant it is produced; a metallic connection must exist from the generator to the end-user; it cannot be stored in large volume; and it remains expensive to produce in small quantities.

Production

All conventional, large-scale electricity production uses the same fundamental technology in which a turbine, propelled by steam, water or gas, is used to drive a generator (Fig. 4.7a). For smaller scale facilities, internal combustion engines or wind-driven blades may be coupled directly to a generator. It is usual to distinguish between the thermal and hydro production of electricity (Fig. 4.7b). The former refers to all supply systems which convert heat into electricity most commonly by using a fuel to produce steam but including systems using geothermal heat, solar radiation and internal combustion engines directly coupled to generators. Hydro-electricity refers to supply systems which gener-ate electricity by means of harnessing the potential energy of the mass of water flowing in a river or the rise and fall of tidal water. In addition, it is useful to

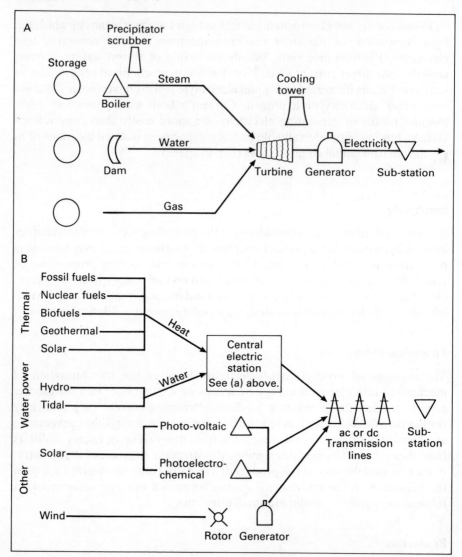

Figure 4.7 Functional components of the electricity industry: (a) central electric station configuration, (b) electricity supply chains

recognize a third category to include what are sometimes referred to as alternative or unconventional electric supply systems based upon wind and, potentially, solar powered photo-electric devices, wave and ocean thermal gradients.

The bulk of the world's electricity is produced by thermal systems using fossil and, to a growing extent, nuclear fuels. Other heat sources such as geothermal are used in only a few parts of the world (e.g. New Zealand, California).

Hydro-electric systems, while still comprising less than 25 per cent of global electrical capacity, are regionally and even nationally important in several countries. Tidal systems, on the other hand, in the mid-1980s were operating at a commercial scale in only France and the USSR.

The size of central electric generating stations and the individual generating sets within them have grown steadily since the beginnings of the industry at the turn of the century although in recent years the trend to ever larger installations appears to have been reversed. The gross capital costs of electricity generating systems have also increased significantly. As the size of plants has increased, new technologies have been adopted (e.g. use of nuclear fuels), more remote hydro sites have been developed and increasingly stringent environmental regulations have been enforced, costs per kW of installed capacity have risen. Accompanying this increase in gross capital requirements there has been a pronounced lengthening of the time required for project approval and construction.

Table 4.7 Components of the cost of electricity production

1. First cost per kW installed capacity	$= \dfrac{\text{Capital costs} + \text{Interest during construction}}{\text{Installed capacity (kW)}}$
Capital costs:	Site acquisition, survey, preparation Fuel (water) storage and handling facilities Combustion and generation equipment Cooling towers, pollution control Substation
2. Cost per kW hour generated	$= \dfrac{\text{Operating costs per kW hour installed}}{\text{(Hours in a year) (Capacity factor)}}$
Operating costs: Fixed: (annual)	Interest on capital Depreciation Taxes
Variable:	Fuel Labour

Table 4.7 summarizes in a simplified way the components of the cost of electricity production and Table 4.8 illustrates the role of these components in determining the production costs in two hypothetical cases. Table 4.8 illustrates the particular importance of four cost components: interest rate and capacity factor for hydro plus conversion efficiency and fuel costs for thermal.

Interest rate The first cost of large central electric stations now runs to billions (10^9) of dollars. The cost of borrowing (interest rate) such large amounts of capital becomes the single dominant element of the first cost per kilowatt of installed capacity (Fig. 4.8). It is clear that old plants, built at lower capital costs or even with capital cost paid off, are able to produce electricity relatively cheaply, providing that their operating efficiency can be maintained.

Table 4.8 Illustrative cost structure of hydro and thermal electricity

	Hydro			*Thermal* (Coal)	
First cost per kW installed		$1,000.00		$800.00	
Cost per kWh generated					
Fixed – per kW installed					
Interest	(10%)	100.00		80.00	
Depreciation	(0.5%)	5.00	(1.5%)	12.00	
Taxes	(0.5%)	5.00		8.00	
Other		3.00		12.00	
		113.00		117.00	

– per kWh generated

$$\frac{113}{8760 \times 0.60^*} = 2.15c \qquad\qquad \frac{117}{8760 \times 0.60^*} = 2.23c$$

Variable			
Negligible		Conversion Efficiency	9600 BTU/kWh (0.8 lb coal)
		Fuel cost	3.18c/lb at 12,000 BTU/lb
		Fuel cost per kWh	$3.18 \times 0.8 = 2.5c$

Total cost per kWh	2.15c	4.73c

* Capacity factor

Capacity factor Given that the installation cost of one unit of productive capacity is so high, the ability of that unit to pay both the carrying charges and principal is a function of how much it produces. One kilowatt of capacity operated throughout the year (i.e. 8760 hours) would produce 8760 kWh and would be said to have operated at 100 per cent capacity factor. In practice, seasonal and diurnal variations in electricity demand coupled, in the case of hydro, with seasonal variations in water supply, mean that many generating plants operate with a capacity factor in the range of 55–65 per cent. However, within a system of plants some will be used for base load purposes and operate at a capacity factor of 85 or more, while those used for peak load and system standby may operate only once or twice a year and have a capacity factor of less than 10. As a general rule, the higher the capacity factor the lower the fixed costs per kWh produced and *vice versa*.

Conversion efficiency The major variable costs in the thermal production of electricity are the amount of fuel required to generate one kilowatt-hour and its price. The amount required is a function of the conversion efficiency of the equipment in use. New conventional plants operate at 35–40 per cent efficiency (which approaches the thermodynamic limit) leading to heat rates of 8500–9500 BTU/kWh. Significant improvements in conversion efficiency (and thus lower fuel requirements) await the full-scale development of new technologies such as fluidized-bed combustion, combined cycle systems and magneto-hydro-dynamics. Conversion efficiency at hydro plants is much higher and the potential energy of the falling water is converted to electricity at efficiencies in excess of 90.

Fuel costs Once the amount of heat required per kWh of electricity generated is set by the equipment in use, the other variable component of the cost of generation is the price of the heat. This translates into the price of fuel which in Table 4.8 is assumed to be coal at $70.00/tonne or 3.18c/lb. It may be seen that at $100.00/tonne fuel costs would increase to over 4c/kWh whereas at $50.00/tonne they would drop to a little over 2c/kWh. By definition, hydro plants do not require fuel so that variable costs are generally negligible except in cases where water-use charges are imposed or penalties are imposed for too rapid drawdown of reservoirs or reduction of flow to downstream users.

Transmission and storage

Unlike other forms of energy which may be transported by a variety of modes, electricity can only be moved by one, a 'power' line. More accurately, long distance transfer of electricity is carried by 'transmission lines' and local movement by 'distribution lines'. The capacity and range of transmission lines is a function of the voltage at which they operate which, in turn, is raised or lowered at transformer substations along the line. In a general way, 'substation' and 'voltage' are functionally equivalent to 'pumping station' and 'pressure' in pipeline terminology.

Although the range over which electricity can be transported has increased steadily, few lines are longer than 1600 km (1000 miles). Greater use of direct-current (dc) lines (requiring converter as well as transformer substations) and developments in insulators and cables will doubtless continue to permit incremental increases in the capacity and range of ac and dc transmission lines. Despite these developments the distance over which electricity can be transported overland is notably less than other forms of commercial energy and only relatively short (less than 100 km) underwater crossings are possible.

Electricity also differs from primary energy sources in that it can only be stored in small amounts. The 'immediacy' of electricity as a form of energy is one of its great attractions; it is also a characteristic which fundamentally influences the functional structure of the industry. In the absence of being able to store electricity itself, electricity suppliers must (a) store the primary energy which can be used to generate electricity and (b) have available generating capacity to meet the peak demands. Storage of the primary energy can be accomplished by storing supplies of fossil fuels and by impounding water in reservoirs either by damming a river system or building a pumped-storage reservoir. In the latter case electricity is used during low demand periods to pump water to a higher elevation where it is stored as potential energy to be run back through the turbines to generate electricity during periods of high demand.

Spatial characteristics

The spatial distribution of generating facilities reflects decisions which take account of the feasibility and relative cost of transporting the primary input energy versus the electricity. Most renewable primary energy sources are not easily transferable with the result that hydro, geothermal, solar and wind based systems have resource-oriented locations. Fossil and nuclear fuels on the other hand are readily transportable over long distances so that the location of thermal plants may be oriented towards resource, intermediate or market locations. In practice, market and intermediate locations predominate. The actual siting of central electric plants in areas selected by the more general location tendencies is made in response to other variables including community acceptance, environmental regulations and safety considerations.

Most readers of this book will be within a few tens of metres of an electricity meter and the end of a distribution line (overhead or underground) that brings electricity to the house or office. In developed economies, this distribution system with its attendant substations reaches to virtually every building whether in urban or rural areas. The distribution system is, in turn, usually linked to a regional or national transmission system (often blatantly visible as a set of giant towers extending across the landscape) linking the generating station to the end user. Linking the producer to the consumer is, however, only one function that a transmission line performs. Another important function is to provide a link between systems in order that reserve capacity may be pooled and overall capital investment reduced. A third function is to reduce operating costs by permitting lower-cost electricity generated in one system to be sold and delivered to another.

Environmental aspects

The environmental implications of the electricity industry are centred upon the production, transmission and distribution links of the supply chain rather than the end-use stage. At the production end it is helpful to distinguish between plants which use fossil and nuclear fuels. For fossil fuel plants the major environmental concerns arise from the disposal of combustion residuals and waste heat. For both coal and oil fired plants, particulate and gaseous residuals (particularly sulphurous compounds) when discharged into the atmosphere can quickly lead to a deterioration of atmospheric quality (at local, regional and even continental scale) and to the deterioration of other environmental systems as the result of acid rain. Because of their large number and widespread distribution (particularly in the northern hemisphere), fossil fuel fired central electric plants collectively are major contributors to the steady increase of carbon dioxide in the atmosphere.

Withdrawal of the large amounts of water required for thermal electric plants may significantly disturb hydrologic regimes whereas the return of large volumes of hot water to the environment can radically change aquatic eco-systems. The use of cooling towers and ponds and of systems which cascade the heat through other uses reduces the temperature of the water and thus its potential impact.

Nuclear plants do not produce particulate or gaseous residuals in the course of normal operation. Instead, highly radioactive spent fuel rods, moderating and cooling fluids or gases are produced. These require the most meticulous disposal procedures, first to ensure the radiational security of the containers in which the waste is packaged and, then, in the underground (or undersea) environment in which they are placed for long-term storage.

The transmission of electricity also results in environmental disturbance as the result of the clearing and maintenance of rights of way and aesthetic degradation as a consequence of the marked visual contrast between the stark geometry of towers and lines and the natural form of most landscapes. There is increasing evidence that the strong electric fields associated with very high voltage lines may have physiological or psychological effects within a short range of the towers.

Organizational structure

Ownership and operating structures in the electric supply industry are varied and complex. Generating, transmission and distribution facilities are owned by public sector utilities at all levels of government (i.e. municipal, state, national), by investor-owned utilities, by co-operatives of users and by industrial users which consume all (or a large proportion) of the electricity they produce in their own plants. In some countries (e.g. USA) regional-scale investor-owned utilities are dominant while in others national (e.g. UK) or state (e.g. Canada, Australia) organizations control the industry.

As noted earlier, there are compelling economic as well as reliability and safety considerations which encourage the physical interconnection of individual electrical supply systems. In order to plan such interconnections and draw up regular and emergency procedures many countries establish agencies to co-ordinate the systems within their own boundaries and, in some cases (e.g. Western Europe, USA–Canada), provide operational links between their systems and those of adjacent countries.

A further characteristic of the organizational structure of the industry is the extent to which it is regulated. In most countries, whether publicly or privately owned, the industry has to acquire licences from the appropriate government agencies for the construction of all facilities and, in many countries, the price structure and the rate of return on capital is subject to some public control. For the most part, the organizations, the capital investment and the production plants of the electricity industry are large-scale leaving little opportunity for small-scale or intermittent producers of electricity to become part of the supply system. However, in some countries (e.g. USA) regulations have recently been introduced to permit the integration of small, independently owned facilities with those of the large, public utilities (Sawyer and Armstrong 1985).

Summary

This chapter reviews the basic characteristics of the primary energy supply systems and electricity. The same framework is used for each in order to provide a basis for comparison between them (Table 4.1). The supply chain concept is illustrated by means of simplified diagrams showing the major facilities required at each stage of the respective supply systems and the linkages between them.

Establishment of the resource base for the fossil and nuclear fuels requires elaborate and expensive exploration programmes (particularly for oil and gas) followed by detailed delineation and assessment in order to determine the quality, magnitude and producibility of each discovered deposit. For most renewable sources, pre-production activities do not involve 'exploration' so much as confirmation of the amount, quality and temporal characteristics of each source (e.g. the head and seasonal variation of river flow or the solar radiation regime).

Following the establishment of the resource base and the decision to commence production, a linked sequence of functions and associated facilities can be identified for each energy source. All the supply chains involve production, processing, transport and storage but the scale and complexity of the required facilities vary between sources. Thus, although the coal chain must be able to handle a large volume of material, processing procedures are relatively simple and mechanical (except at the combustion end) and environmental consequences are rarely hazardous (again, except at the combustion end). The nuclear chain, on the other hand, deals with small volumes of material in a very complex manner with potentially extreme environmental hazards.

Coal and hydrocarbons may occur relatively close together in some major sedimentary basins (e.g. in Alberta) but generally the spatial association of each with differing sedimentary environments results in discrete distribution patterns at local and regional scales. The distribution of renewable resources results from the operation of widely differing variables including geological conditions (e.g. geothermal), climate, soil and agricultural practices (e.g. biomass) and hydrologic, geomorphic and oceanographic conditions (e.g. waterpower). The four major mineral fuels are not only highly concentrated energy sources but are spatially concentrated and in comparison with most renewable sources have low energy content and extend over large areas (e.g. forests). Because of their high energy content and physical form primary mineral fuels have a high degree of transferability in comparison with renewable sources. The combination of spatial extent and low transferability lead to dispersed patterns of production for most renewables (excluding hydro and geothermal) making it difficult for them to meet the highly concentrated energy needs of urban societies.

All stages of the supply chain of the primary energy sources have environmental implications. Some may be relatively slight, localized and benign, others may be major, hemispheric or even global in extent and hazardous. The more notable impacts are sketched for each link in each supply chain, distinguishing between the environmental disturbance caused by each activity and the effects of residuals and their disposal.

The organizational structure of the industries which mobilize and market the respective energy sources may be summarized in terms of ownership, integration and spatial scale of operation. The resource bases of all the major commercial energy sources generally are publicly owned by national or state governments. Outside the Centrally Planned Economies, operating companies acquire exploration and production rights by lease or licence. Ownership of such renewables as solar radiation or wind is not clearly established and, in many developing countries, jurisdiction over forested areas or agricultural wastes is uncertain and regulation of their use for energy purposes is difficult.

Ownership of the production and downstream links of the respective supply chains is shared by both public and private organizations, the particular mix varying between the different energy sources and over time and place. For example, the coal industry outside the Centrally Planned Economies is generally privately owned (except in the United Kingdom and South Africa) whereas in the oil industry public ownership became increasingly important in the 1970s in almost all countries except the USA. The electricity supply industry, on the other hand, has had a large element of public ownership since its early days.

The degree of vertical integration also varies between sectors being least in the coal and uranium industries and greatest in the oil and electricity sectors. The natural gas industry has less vertical integration than the oil in part because of the larger number of independent and smaller scale companies. Horizontal integration grew rapidly after the early 1970s as companies with a base in one energy sector (e.g. oil) acquired interests in others (e.g. coal and uranium) to become comprehensive 'energy' companies.

Spatially the operations of energy supply corporations vary from the global range of the multi-national companies, the international scope of some of the coal and uranium organizations to the national, regional and local span of electric utilities. The latter may be part of an international organization to permit peak load sharing and operation during emergency conditions.

CHAPTER 5

Geography of supply

This chapter reviews the spatial structure of commercial energy supply. By limiting the coverage to commercial energy (fuels and electricity traded at a global and international scale), the major energy supplies for the subsistence economies of the developing world are excluded. These supplies are largely of biomass origin and are either gathered on a family or village basis or they are traded on a small and local scale, often in a semi-formal manner. Important as these supplies are to almost half the world's population they are estimated to constitute only 10 per cent of the total world energy supply (Wood and Baldwin 1985). Although the literature on traditional fuel supply systems is growing, there are still insufficient reliable data to permit comparative analysis at international and national scales.

At the most generalized spatial scale, the Developed Market Economies (DME) and the Centrally Planned Economies (CPE) each produce a similar proportion of the world's total primary commercial energy and of liquid fuels but differ in the relative proportion of the world total of solid fuels, gas and, particularly, primary electricity (Table 5.1). The DME, however, are large net importers whereas the CPE are net exporters, The Developing Economies (DE) produce a little more than one quarter of the total commercial energy but, within that, almost a half of the liquid fuels,which largely accounts for their strong net export position.

Although the level of generalization of Table 5.1 provides an informative introductory background, it is too broad to serve as the basis for a review of the geography of energy supply. The following sections deal with each of the major commercial energy sources in turn, paying attention to individual links in the respective supply chains. The data are reviewed in an evolutionary way, starting at mid-century and coming as close to the present as available data permit. The spatial scale ranges from global, through major economic systems to individual countries.

Table 5.1 Total primary energy: World production, trade and consumption, 1983

	Production					Exports	Imports	Consumption				
	Total	Solids	Liquids	Gas	Elec.*			Total	Solids	Liquids	Gas	Elec.*
World 10¹⁵ J	263	79	118	54	11	79	81	250	81	104	55	10
%	100	30	45	21	4			100	32	42	22	4
						Percentage of world						
Developed Market	38	41	26	50	64	27	69	50	37	56	53	7
Centrally Planned	36	52	27	39	18	18	10	36	52	25	38	10
Developing	27	8	47	11	18	56	21	14	11	20	9	20

* Hydro, geothermal, nuclear
Source: UN 1985, Table 3

109

Oil

Reserves

The macro-geography of estimated proved reserves of conventional oil in the twentieth century displays two major patterns, the first centred on the USA and the second on the Persian Gulf states. At the turn of the century and for approximately the next 40 years, the majority of the world's known reserves were in the USA with first the USSR and then Iran, Iraq and Venezuela

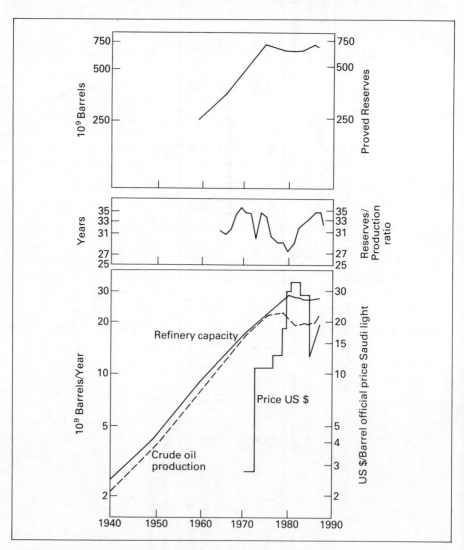

Figure 5.1 Trends in the world oil industry, 1940–86
Sources: see Tables 5.2–5.7

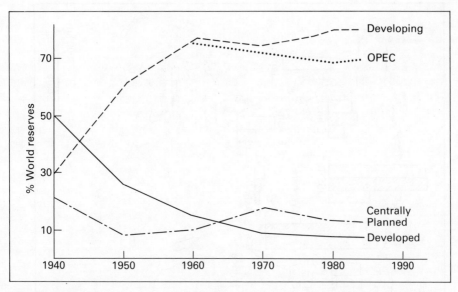

Figure 5.2 Estimated proved crude oil reserves: percentage of world by major economic groups, 1940–86
Sources: see Table 5.2

gradually increasing their share. By 1940, with the discovery of the giant Abqaiq field in Saudi Arabia, the stage was set for a major change. By 1950 the USA was replaced by Saudi Arabia and adjacent Persian Gulf states as the focus of global reserves. As the known total increased (Fig. 5.1) and despite major discoveries in the Developed Market and Centrally Planned Economies, the Developing Economies maintained their dominance with over three-quarters of estimated proved reserves (Tables 5.2 and 5.3). Within these economies there is a pronounced concentration in a small number of countries (mainly members of OPEC), leaving the majority with little or no known oil reserves.

While the global pattern of the distribution of reserves has remained remarkably stable for almost 50 years, there have been regional and local changes of general significance to the industry as a whole as well as to individual national and regional economies. Of general significance has been the growth of the proportion of reserves located offshore. Starting in the shallow waters of areas such as Lake Maracaibo and the Mississippi delta, reserves have now been established in many offshore locations and under ever deeper water. In the context of national development, discoveries in North and West Africa have provided great opportunities for economic growth while those in the North Sea have provided the UK and Norway with an unexpected degree of independence in their energy economies. Within countries, the pattern of discoveries in the USSR and Canada has major regional implications for Siberia and Alberta respectively. This brief discussion of reserves refers to

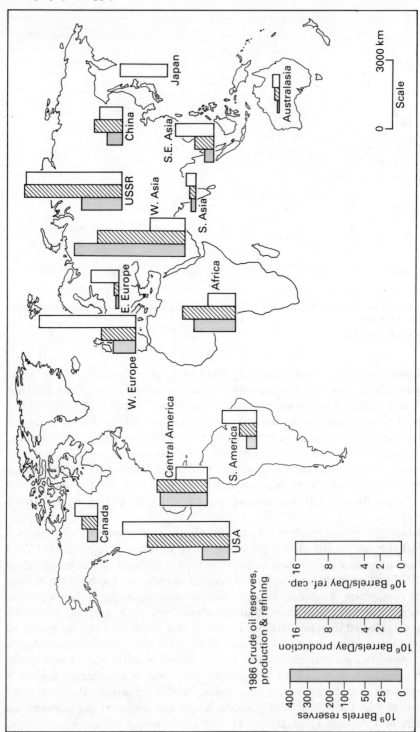

Figure 5.3 Crude oil reserves, production and refining: World, 1986
Source: British Petroleum 1987

Table 5.2 Estimated proved oil reserves: major economic groups and regions, 1940–86

	1940	1950	1960	1970	1980	1986
			10^9 brls			
World	34	103	266	621	649	700
			Percentage of world			
Developed Market	50	26	14	10	9	9
of which						
N. America	50	26	14	9	5	5
W. Europe				1	4	4
Centrally Planned	21	8	10	16	13	12
of which						
USSR	21	8	9	14	10	9
China	—	—	—	2	3	3
Developing	29	66	76	74	78	79
of which						
OPEC	—	—	74	72	68	68
of which						
P. Gulf	—	—	61	60	56	57

Sources: British Petroleum 1986 and 1987; *International Petroleum Encyclopedia* 1981

Table 5.3 Estimated proved oil reserves and reserve/production ratio: major economic groups and regions, 1986

	Estimated proved reserves 10^9 brls	Country	Estimated proved reserves 10^9 brls	Reserve/ production ratio Years
World	703		703	33.5
Developed Market	60	Saudia Arabia	167	90
of which		Kuwait	92	over 100
N. America	40	USSR	59	13
W. Europe	18	Iran	49	71
Oceania	2	Mexico	55	56
		Iraq	47	75
Centrally Planned	79	USA	33	9
of which		Abu Dhabi	31	81
USSR	59	Venezuela	25	39
		Libya	21	55
Developing	564	China	18	18
of which		Nigeria	16	30
OPEC	478	Norway	11	31
of which		Algeria	9	27
P. Gulf	395	Indonesia	8	17
N. Africa	30	Canada	8	12
W. Africa	17			
L. America	27			
S. E. Asia	9			

Source: British Petroleum 1987

Table 5.4 Estimated proved oil reserves and exploration drills
active: major economic groups and selected
regions, 1984

	Estimated proved reserves	Exploration drills active
	Percentage of world 1984	
Developed Market of which	9	75
N. America	5	69
W. Europe	4	6
Developing of which	79	25
OPEC	68	8
Centrally Planned	12	no data

Source: McCaslin 1985

conventional oil only. If heavy oil, bitumen and kerogen are considered, the very limited data that are available reveal a quite different pattern in which Canada, the USA and Venezuela are the dominant countries.

The macro-pattern of reserves which has persisted during the second half of the twentieth century will continue to prevail unless, of course, large discoveries are made in new areas. Such discoveries will only result from exploration, the current distribution of which outside the USSR and Eastern Europe is shown in Table 5.4. By far the greatest exploration effort is concentrated in North America with 69 per cent of active drills, less than 10 per cent in OPEC (and less than 5 per cent in the Persian Gulf states) but nearly 20 per cent in other developing countries. Thus the geography of oil exploration outside the USSR is quite different from that of reserves, being concentrated in consuming countries either with declining reserves (e.g. USA) or with promising, but little explored, geological conditions (e.g. China).

Production

Trends in the macro-geography of oil production reflect that of estimated reserves with two notable differences: a shift in the crossover dates of the relative importance of the three major economic groups (DME, CPE, DE) and, since 1976, a convergence of their relative importance (Fig. 5.4). Production from the Developing and Centrally Planned Economies overtook that in the Developed Market Economies in 1958 and 1974 respectively, some 15–20 years after the reserves had been established in those areas.

Since 1976, however, the leading position of the developing countries has weakened somewhat, first as the result of the relative decline of production from OPEC generally and then starting in 1980, from the Persian Gulf states particularly (Table 5.5). The erratic production of the three major Persian Gulf

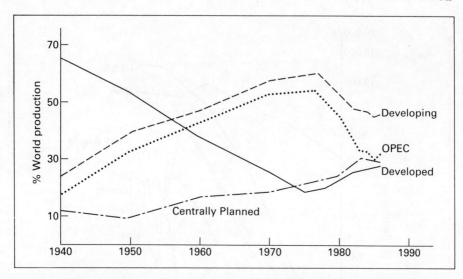

Figure 5.4 Production of crude oil: percentage of world by major economic groups, 1940–86
Sources: see Table 5.5

Table 5.5 Crude oil production: World, major economic groups and selected regions, 1940–86

	1940	1950	1960	1970	1980	1986
World (10³ brls/d)	5,757	10,428	21,088	45,060	59,423	54,567
			Percentage of world			
Developed Market	66	53	37	25	22	28
of which						
N. America	65	53	36	24	16	19
W. Europe	—	—	1	1	4	7
Centrally Planned	10	8	16	18	25	27
of which						
USSR	10	7	14	16	20	22
Developing	24	39	47	57	53	45
of which						
OPEC	17	33	41	52	45	31
of which						
P. Gulf	5	17	25	29	31	19

Sources: McCaslin 1986; British Petroleum 1986

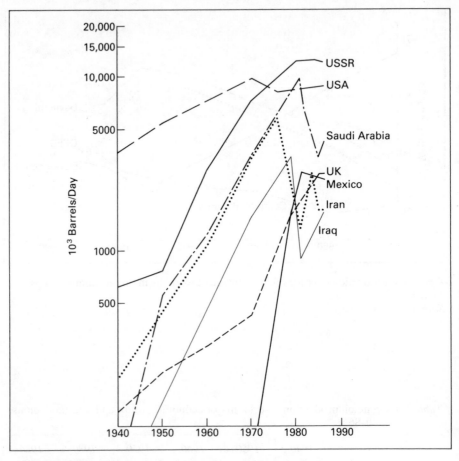

Figure 5.5 Production of crude oil: selected countries, 1940–86
Sources: see Table 5.5

states since 1980 is evident from Figure 5.5. The performance of the industry in Iran and Iraq is the result of military hostilities whereas the decline in Saudi Arabian production between 1981 and 1985 arose from a policy decision. Production was cut in order to maintain prices in the face of falling demand, full-scale output from other OPEC producers and additional production from newly discovered fields in, for example, the UK and Mexico. The 1986 upturn in production is evidence of a further change in policy attended by a dramatic decrease in price (Fig. 5.1).

The Centrally Planned Economies continue to increase their share of world output but, in absolute terms, production in the USSR has stabilized at 12 million barrels daily despite policy expectations of increased output. This stabilization appears to be caused by disappointing exploration results and delays in applying secondary and tertiary recovery techniques to ageing fields.

In the USA, on the other hand, production peaked in 1970 at nearly 10 million barrels daily, fell to almost 8 million in 1976 but has shown a marginally rising trend since. This increase has been accomplished by the large scale application of enhanced recovery techniques to mainly old fields. The advanced stage of production of most fields in the USA is indicated by an average production per well of only 20 barrels a day compared with 150 in the USSR, 320 in the North Sea fields and 5500 in the Persian Gulf states (British Petroleum 1979).

In keeping with the growth of reserves, the proportion of world crude production from offshore locations has steadily increased. As early as 1970, 15 per cent of total production was from offshore wells and by the mid-1980s the proportion was approximately 30 per cent (American Petroleum Institute 1986). In the more remote and deeper water areas, offshore oil is relatively high cost with the result that during periods of depressed prices, production from such reserves may be curtailed.

In terms of the future it is important to recognize the increasing production of heavy oil in many parts of the world (but particularly Venezuela) and the output of 'synthetic' oil from the bitumen in Canadian tar-sand deposits. As and when the price of conventional oil resumes its upward trend these resources will become increasingly attractive and, in the long-term, could alter the geography of oil production significantly.

Processing

The basic distribution of oil refining, the second stage of the supply chain, is shown in Table 5.6 and Figure 5.6. In 1940, the Developed Economies both

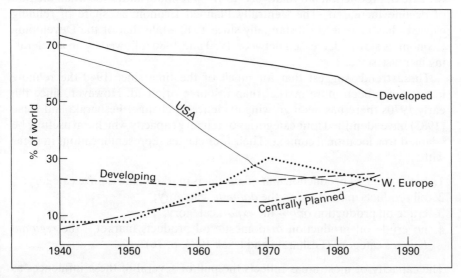

Figure 5.6 Oil refining capacity: per cent of world by major economic group, 1940–86
Sources: see Table 5.6

Table 5.6 Oil refining capacity: World, major economic groups and selected regions, 1940–86

	1940*	1950	1960	1970	1980	1986
World (10^3 brls/d)	6,608	11,647	24,470	49,212	79,515	74,753
			Percentage of world			
Developed Market of which	78	67	66	64	59	52
N. America	71	60	47	27	25	23
W. Europe	7	7	16	30	25	21
Centrally Planned of which	N/A	10	14	14	19	23
USSR		8	11	11	14	16
Developing of which	22	22	20	21	22	25
OPEC of which	9	11	10	8	6	8
P. Gulf	6	7	5	4	3	4

* 1940: Excluding centrally planned   N/A = Not available
Sources: McCaslin 1986; British Petroleum 1986

produced and processed the majority of world oil (66 per cent and 72 per cent respectively). Subsequently, their share in both has declined, dramatically in the case of production but relatively slowly in refining until 1980 when the rate of decline increased. The Centrally Planned Economies' share of refining capacity has increased substantially since 1940 while that of the Developing Economies slowly decreased between 1950 and 1980 followed by an accelerating increase since.

These trends suggest that for much of the time since 1940 the refining industry has been more market than resource oriented. However, since the early 1970s there has been growing evidence of change. Fesharaki and Isaac (1983) have identified four categories of refinery capacity which can usefully be adapted to a locational context. Their four classes represent locations in areas with:

1. both crude oil production *and* oil products market – the *captive* category;
2. oil products market *only* – the *domestic* category;
3. crude oil production *only* – the *export* category;
4. no crude oil production *or* domestic oil products market – the *regional balance* category (Fesharaki and Isaak 1983, p. 76)

The capacity of most areas reflects the pull of several of these influences as Table 5.7 indicates.

Viewed in these terms, the rapid growth of world refining capacity from 1940

Table 5.7 Oil refining capacity by category: World, selected regions, and countries, 1980

	Total capacity 10^3 brls/d	Percentage of total			Regional balance
		Captive	Domestic	Export	
World	81,936	35	39	8	18
N. America	20,631	49	40	—	15
W. Europe	20,190	10	56	—	34
Oceania	861	42	40	—	18
Japan	5,662	—	88	—	12
E. Europe	13,724	67	13	20	—
Africa	2,179	32	26	27	15
L. America	8,674	26	23	20	31
Mexico	1,394	65	—	35	—
Venezuela	1,349	21	—	79	—
N. Antilles					
Far East					
Indonesia	515	71	—	29	—
Singapore	1,049	—	36	—	64

Source: Fesharaki and Isaac 1983

to 1980 was dominated by the captive type of location (e.g. USA), followed by the balance type (e.g. Netherlands Antilles) with a few examples of the export type (e.g. Abadan in Iran). By 1970, the captive type had declined in relative importance (with the exception of the USSR) and the domestic (e.g. Western Europe and Japan) and balance (e.g. US Virgin Islands and Singapore) types increased. By 1985 the locational balance appeared to be shifting again as the domestic and balance types were bearing the brunt of closure (e.g. Japan and Netherlands Antilles) while the captive and export types were experiencing some growth (e.g. Persian Gulf).

The developments in the 1980s were the result of changes in market conditions and industry organization. Markets for petroleum products have shown general weakness since 1980 and shifted markedly from fuel oils towards gasoline. The reduced demand resulted in excess refinery capacity in most of the major market areas and the declining market for fuel oils accompanied by the increasing proportion of heavy crude, means that viable refineries must have considerable reforming and cracking capability. Organizationally, the world petroleum industry has dramatically changed. The fully integrated, multi-national corporations with control of the complete supply chain from exploration to retailing have, to a considerable degree, lost control of their supplies of crude to the government controlled agencies of producer countries. These agencies are themselves integrating downstream into refining and petrochemicals.

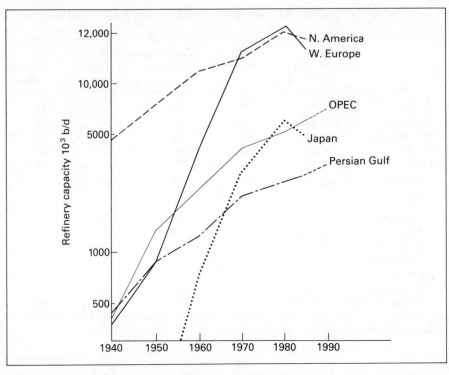

Figure 5.7 Oil refining capacity: selected regions, 1940–86
Sources: see Table 5.7

The downstream developments of the producer country companies may be accomplished by purchasing existing operations in market areas which, in effect, results in continuing the prevailing spatial structure. Another strategy is to build export refineries in the home country, thus changing the geography of refining. In response, the multi-national corporations have closed a number of the older, smaller and least technically sophisticated refineries, and upgraded others. For example, in Canada eleven refineries were closed between 1980 and 1986 and eight of the remaining twenty-eight were undergoing upgrading in 1986 (*Oilweek* 1986b). Capital intensive upgrading provides the refineries with increased technical capability to match their output to changing market requirements and thus compensate for the flexibility that they were formerly able to achieve by relatively free access to a wide range of crudes of different quality.

Trade

The macro-geography of production and processing provides the framework within which international trade in crude and products evolves. The trade is predominantly in crude oil both in terms of proportion of production traded

Table 5.8 Exports of crude oil, gasoline and fuel oil as percentage of production: World, 1950–85

	Crude oil			Gasoline*			Fuel oil**		
	Prod. 10^6 tonnes (1)	*Export* (2)	*Per cent* 2 of 1 (3)	*Prod.* 10^6 tonnes (1)	*Export* (2)	*Per cent* 2 of 1 (3)	*Prod.* 10^6 tonnes (1)	*Export* (2)	*Per cent* 2 of 1 (3)
1950	520	141	27	160	22	14	272	82	25
1960	1052	381	36	287	31	11	552	149	25
1970	2270	1162	52	495	39	6	1281	284	21
1980	2979	1483	50	650	47	7	1628	261	16
1984	2699	1110	41	667	59	9	1451	290	20
1985	2660	1020	38	670	61	9	1414	284	20

* Aviation and motor   ** Middle distillate and residual
Sources: UN various years

and absolute volume (Table 5.8). As the output of crude increased after World War II, the proportion of production which was exported also increased rapidly, reaching a maximum of 56 per cent in the late 1970s and then declining to approximately 40 per cent by 1986. Conversely, the proportion of the total output of products that was traded steadily declined from 1950 to 1980, followed by an increase. These trends suggest a lagged response to the changing geography of refineries discussed above.

Although still providing approximately three-quarters of the world's exports of crude, the relative role of the Developing Economies has declined since 1950 and that of the CPE (USSR) and DME (UK and Norway) increased. Among the Developing Economies, a significant decrease of the role of OPEC and, within that group, of the Persian Gulf states occurred in the early 1980s.

Imports of crude oil are concentrated in the Developed Economies which received over 75 per cent of the world's imports during the early and mid-seventies declining to 65 per cent by 1983 (Table 5.9). Within these economies the role of Western Europe, Japan and the USA is particularly important. In 1970, Western Europe was the destination for over half of the world's crude imports, but by the early eighties the proportion had declined to approximately one-third. Between 1970 and 1980, on the other hand, the proportion of world imports taken by the USA increased from 8 to 19 per cent which, in absolute terms, represents an increase of 165 million tonnes (from 96 to 269 million tonnes). Subsequently crude imports have declined to approximately 160 million tonnes per annum. The overall relative decrease in the importance of the Developed Economies as importers since the seventies has been accompanied by an increase in that of the Centrally Planned (particularly Eastern Europe) and the Developing Economies. Within the latter, newly industrializing and 'balance' refining countries (e.g. Brazil, and Singapore, respectively) have become increasingly large importers.

Table 5.9 Imports and exports of crude oil: major economic groups, 1950–85

| | Imports | | | | | Exports | | | | |
	1950	1960	1970	1980	1985	1950	1960	1970	1980	1985
Developed Market										
10^6 tonnes	78	251	897	1076	813	6	8	35	92	155
%	56	73	77	70	72	5	2	3	6	15
Centrally Planned										
10^6 tonnes	1	8	43	112	103	—	18	67	134	146
%	1	2	4	7	9	—	5	6	9	14
Developing										
10^6 tonnes	60	94	230	346	214	134	355	1060	1257	720
%	43	25	20	23	19	95	93	91	85	71

Sources: UN various years

Trade in petroleum products is more complex and less easily documented than that in crude. Table 5.10 illustrates the gross trading patterns of gasolines and fuel oils. Even though they have most of the refining capacity, the DME are the major importers. Among the largest consumers in the DME, Western European countries steadily increased their dominant share of world product imports until 1970 when the proportion of gasoline started to decline although that of fuel oil continued to increase. The USA share of both product groups also increased up to 1970 but then, in contrast to Western Europe, the share of fuel oil started to decline while that of gasoline increased. These trends reflect the influence of such variables as changing product-use in response to price increases and the growing market share of export refineries in the oil-producing countries.

Exports of products necessarily originate from those areas with balance and export refineries and, as a result, all three major economic systems export to some extent. Among the DME, Western Europe has steadily increased its share of world gasoline exports while that of the Developing Economies has been declining. The trend has been the same for fuel oil, although the Developing country exporters still have the largest share of total world product exports. As export refinery capacity grows in producer countries (particularly in the Persian Gulf states), these trends are likely to change.

The direction of trade flows in crude and products is shown in Tables 5.11 and 5.12 for 1985. Notwithstanding the relative decline in importance noted above, the heart of the global export trade in crude oil remains the Persian Gulf states with very large flows to Western Europe, Japan and South-east Asia with lesser amounts to Latin America, North America and Africa. Other major flows include those from the Caribbean to North America and Europe, from Indonesia to Japan and the USA, from the USSR to Eastern and Western

Table 5.10 Imports and exports of gasoline and fuel oil: major economic groups, 1950–85

	Imports					Exports				
	1950	1960	1970	1980	1985	1950	1960	1970	1980	1985
Motor gasoline										
Developed Market										
10^6 tonnes	12	14	20	25	38	4	10	13	23	29
%	55	45	65	69	70	18	32	33	49	49
Centrally Planned										
10^6 tonnes	2	5	3	2	8	2	6	5	8	19
%	9	16	10	6	15	9	19	13	17	31
Developing										
10^6 tonnes	8	11	8	9	8	16	15	21	16	12
%	36	35	26	25	15	73	49	54	38	20
Fuel oil										
Developed Market										
10^6 tonnes	41	98	228	174	196	10	35	88	76	114
%	60	72	84	71	80	12	23	31	29	40
Centrally Planned										
10^6 tonnes	1	2	5	5	6	1	15	27	44	62
%	2	2	2	2	2	1	10	10	17	22
Developing										
10^6 tonnes	26	34	41	67	44	72	99	169	138	102
%	38	26	15	27	18	88	67	59	53	38

Sources: UN various years

Europe and Cuba, and, most recently, from the North Sea (UK and Norway) to Western Europe and North America.

In 1985, the volume of international trade in petroleum products was only a little over 30 per cent of that in crude oil. As with crude, the Persian Gulf countries were the major exporters followed closely by the Caribbean states of Latin America and the USSR. Although net product importers, both the USA and Western Europe export significant volumes. South-east Asia is also an origin and destination of product flows reflecting, in part, competition between the balance refineries of the region (e.g., Singapore) and the low cost export plants of the Persian Gulf.

Table 5.11 Trade in crude oil: selected regions and countries, 1985

Crude oil 10³ brls/d

	Total exports	W. Eur.	Japan	USA	E. Eur.	S. E. Asia	L. Amer.	Africa	Canada	Other
					To:					
P. Gulf	8205	2650	2490	255	375	1195	635	300	35	625
L. America	2470	615	145	1215	—	70	275	40	110	—
USSR	2270	640	—	—	1375	30	150	5	—	1445
N. Africa	2045	1510	45	105	180	—	75	10	35	265
W. Africa	1730	945	—	490	—	10	75	80	10	25
S. E. Asia	1005	5	570	295	—	—	170	—	—	105
China	585	15	220	35	—	315	—	—	—	—
Canada	470	—	—	470	—	—	—	—	—	—
W. Europe	380	—	—	305	5	—	10	—	60	10
Other	410	25	255	60	—	30	—	—	20	20
Total		6405	3505	3195	1935	1650	1465	435	270	2495

Source: British Petroleum, personal communication

124

Table 5.12 Trade in petroleum products: selected regions and countries, 1985

	Total exports	Petroleum products 10³ brls/d To:								
		W. Eur.	USA	Japan	S. E. Asia	L. Amer.	Africa	E. Eur.	S. Asia	Other
P. Gulf	1135	370	170	225	195	—	50	—	65	60
L. America	1095	190	790	—	—	145	15	—	—	100
USSR	925	585	—	10	25	135	—	115	45	—
USA	565	160	—	105	65	135	10	5	—	85
S. E. Asia	550	45	115	245	—	—	30	—	25	135
W. Europe	425	—	165	5	5	5	65	30	—	275
E. Europe	420	370	45	—	—	—	—	—	—	5
N. Africa	370	215	100	10	—	25	10	—	—	10
Canada	285	5	280	—	—	—	—	—	—	—
Other	290	5	50	90	120	—	15	—	—	10
Total imports		1960	1715	690	415	355	195	155	135	680

Source: British Petroleum, personal communication

Natural gas

Reserves

Despite their co-existence in many deposits, the geography of natural gas supply differs from oil in a number of ways. First, although much gas is associated with oil, a growing proportion of reserves is found in gas-only reservoirs. Second, not all gas produced is marketed. Third, unlike oil, not all gas has to be processed (and some only requires very little processing) before it can be used. Finally, the cost structure of natural gas transport constrains inter-continental trade.

Estimated proved world reserves of gas are still increasing significantly with recent major additions in Persian Gulf states (Iran and Qatar), the USSR, Australia and Norway. In 1985, the USSR and Persian Gulf states together contained almost three-quarters of the world reserves (Table 5.13) with most of the remainder in North America, Western Europe, North Africa and South-east Asia.

Table 5.13 Estimated proved reserves of natural gas: World, major economic groups and regions, 1975 and 1986

	1975	*1986*
	10^{12} *cu metres*	
World	61	98
	Percentage of world	
Developed Market	22	15
of which		
N. America	12	8
W. Europe	8	6
Centrally Planned	39	45
of which		
USSR	37	43
China	1	1
E. Europe	1	1
Developing	43	40
of which		
OPEC	36	32
of which		
P. Gulf	24	25

Sources: *International Petroleum Encyclopedia* 1976; British Petroleum 1987

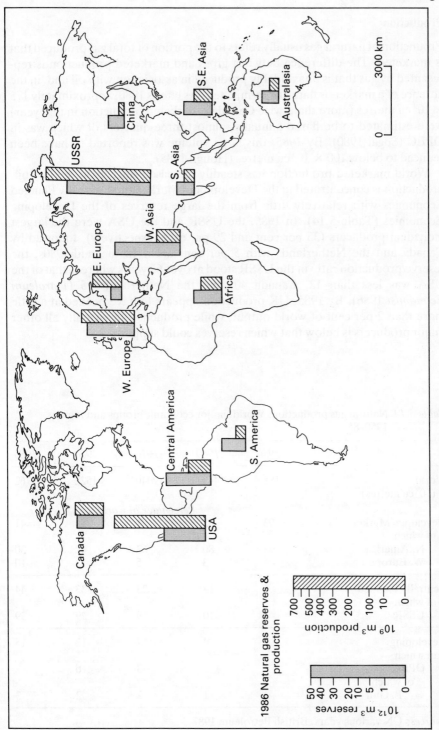

Figure 5.8 Natural gas, reserves and production: World, 1986
Source: British Petroleum 1987

Production

Production of natural gas usually refers to the portion of total gas produced that is marketed. The difference between gross and marketed production is represented by gas that is unavoidably produced in association with oil and, in the absence of a market, is flared or re-injected. As late as 1980, approximately 175 $\times 10^9$ cu metres (more than twice Canada's marketed production in that year) were estimated to be flared annually, almost three-quarters of which was in OPEC (Segal 1980). By 1985, this large volume was reported to have been reduced to below 100×10^9 cu metres (Hough 1985).

World marketed production has steadily increased since 1950. Unlike oil, production is concentrated in the Development Market and Centrally Planned Economies with relatively little from the large reserves of the Developing Economies (Table 5.14). In 1985, the USSR and the USA were the largest individual producers (35 per cent and 29 per cent respectively), followed by Canada and the Netherlands with 5 per cent each. In the mid-1980s, the reserve/production ratio in the USSR stood at over 65 years whereas that of the USA was less than 12, Canada 40, and the Netherlands 15 (*Petroleum Economist* 1986). By 1985, UK production appears to have peaked at a little more than 2 per cent of world output, while production in virtually all other major producers is below that which reserves could support.

Table 5.14 Natural gas production: World, major economic groups and regions, 1950–85

	1950	1960	1970	1980	1986
World (10⁹ cu metres)	185	449	1040	1522	1762
		Percentage of world			
Developed Market	93	83	71	55	41
of which					
N. America	93	80	63	42	30
W. Europe	—	3	5	13	10
Centrally Planned	5	13	23	33	44
of which					
USSR	3	10	19	29	39
Developing	2	4	7	12	15
of which					
OPEC	1	2	3	6	7
of which					
P. Gulf	—	1	2	2	3

Sources: UN various years; British Petroleum 1987

Processing

Unlike crude oil, only a small proportion of natural gas is processed to any extent. Of the three major types of processing applied to natural gas (desulphurization, production of natural gas liquids (NGL) and liquefaction) the production of NGL such as ethane, propane, butane (valued both as fuels and petro-chemical feedstock) is the most sophisticated type. NGL production is concentrated in the USA with almost 50 per cent of world capacity (excluding the CPE) followed by Canada, Mexico, Algeria and Libya (Table 5.15). Additional capacity is under construction or planned in the Persian Gulf states (particularly Saudi Arabia) and, although comprehensive data are not available, large processing facilities are reported to be under construction in the USSR (*Oil and Gas Journal* 1986).

Table 5.15 Natural gas processing capacity: World, major economic groups and regions, 1985

	Capacity 10^9 cu ft/day	Production 10^6 gal/day
World	129	116
	Percentage of world	
Developed Market	78	66
of which		
N. America	68	59
W. Europe	8	3
Centrally Planned	Data not available	
Developing	22	34
of which		
OPEC	15	16
of which		
P. Gulf	7	3
N. Africa	4	10

Source: McCaslin 1987

Trade

The proportion of natural gas production entering international trade increased rapidly after 1960 but, since 1980, has stabilized at a level considerably below that of crude oil (Table 5.16). The basic reason for this more limited trade in gas is the high capital cost of the necessary transport systems arising largely from the low energy density of gas in comparison with oil. In 1984, approximately three-quarters of the gas entering international trade moved in pipelines and the remainder by Liquefied Natural Gas (LNG) tankers (Table 5.17).

Table 5.16 Export of natural gas as percentage of production:
World, 1950–85

	Production (1)	Export (2)	Percentage 2/1 × 100
	10^9 cu metres		
1950	185	1	1
1960	449	5	1
1970	1040	45	4
1980	1522	202	13
1986	1762	221	13

Sources: UN various years; British Petroleum 1987

In 1985, the first tier of exporters (by pipeline) was led by the USSR followed by the Netherlands, Norway and Canada. The very large reserve base of the USSR coupled with its need to earn hard currency suggests that it will remain in this position at least for the remainder of the century. Both Norway and Canada also have substantial reserves in relation to their own needs and are also likely to be able to sustain large exports. Norwegian gas will probably replace the declining shipments of Dutch gas in the Western European markets. Canadian exports to the USA declined in the early eighties as the result of difficulties in competing in the US market rather than reserve restrictions.

Exports by tankers are on a smaller scale and dominated by Indonesia and Algeria. Although the tanker-borne trade has increased substantially, the high expectations of the mid-seventies have not fully materialized as a consequence of the overall softening of energy markets, the high capital costs of the LNG supply chain and, in 1986, the dramatic decline in the price of oil.

Western Europe and Japan together receive three-quarters of the world imports with the majority of the remainder going to the USA. Western and Southern Europe are supplied predominantly by pipeline (North Sea, USSR and North Africa) with some LNG from Algeria and Libya. Japan, on the other

Table 5.17 Natural gas exports: major countries, 1986

By pipeline	10^9 cu metres	By tanker	10^9 cu metres
USSR	77	Indonesia	21
Netherlands	30	Algeria	12
Norway	27	Brunei	7
Canada	21	Malaysia	7
Others	14	Others	5
Total	169		52

Source: British Petroleum 1987

hand, is wholly supplied by tankers mainly from South-east Asia with smaller amounts from Abu Dhabi and Alaska. Contracts have been let to add Australia to the list of LNG suppliers to Japan by 1990.

In sum, the international trade in natural gas is divided into three major spatial systems: the Pacific system focusing on Japan as the buyer, the West European system with multiple buyers and supplies from the USSR and North Africa in addition to the indigenous supplies from the North Sea, and the largely self-sufficient North American system built around the large production and consumption of the USA supplemented by imports from Canada and, to a lesser extent, Mexico.

Coal

Reserves

In terms of contained energy, the estimated 'proved' reserves of coal in the world are significantly larger than other fossil fuels and the reserve to production ratio correspondingly greater. The reserves are distributed roughly equally between the DME and CPE with only a small proportion in the DE (Table 5.18). The USA and USSR both contain approximately one-quarter of the world's reserves with 50–55 per cent hard coal in each. The quality and accessibility of the US reserves, however, is generally greater than those of the USSR. Western and Eastern Europe and China each have 10–11 per cent of global reserves made up of less than 50 per cent hard coal in Europe but a reported 100 per cent in China.

Production

World coal production has increased steadily since 1950, rising by 25 per cent between 1970 and 1980 and a further 16 per cent by 1985 (Table 5.19). The USA, USSR and China between them produced almost 60 per cent of global production in 1985 with China recording the most rapid increase in the 1980–85 period. Exporting countries such as Australia, Canada and South Africa, with smaller absolute production, have nevertheless shown the greatest increase in output since 1980. Production in Western Europe on the other hand, which amounted to almost one-half of the world total in 1938, had declined to only 10 per cent by 1985. In the 15-year period 1970–85, hard coal production in Western Europe was reduced by almost 100 million tonnes although, in the same period, soft coal output increased by 40 million tonnes. Reflecting increasing mining depths and rising costs, the major decline in hard coal production occurred in the UK and West Germany.

Over two thirds of the small total production in the Developing Economies comes from one country, India. With limited hydrocarbon resources, India is pursuing a policy of rapidly increasing coal output as may be seen from the 40

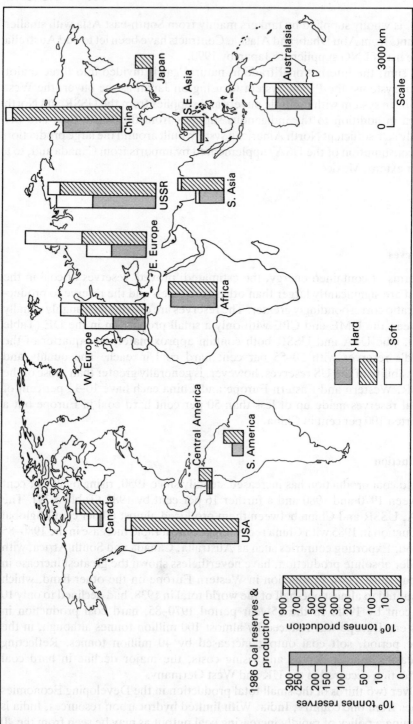

Figure 5.9 Coal, reserves and production: World, 1986
Source: British Petroleum 1987

Table 5.18 Estimated proved reserves of coal: World, major economic groups and selected countries, 1986

	10^9 tonnes			
	Hard*	Soft**	Total	% World
World	579	439	1018	100
Developed Market	257	234	491	48
of which				
N. America	136	135	271	27
of which				
Canada	4	3	7	<0.5
USA	132	132	264	26
W. Europe	35	61	96	9
of which				
W. Germany	24	35	59	6
UK	9	1	10	1
Oceania	28	38	66	6
of which				
Australia	27	38	65	6
S. Africa	58	—	58	6

	10^9 tonnes			
	Hard*	Soft**	Total	% World
Centrally Planned	297	196	493	48
of which				
China	156	14	170	17
USSR	109	136	245	24
E. Europe	32	46	78	8
of which				
E. Germany	—	21	21	2
Poland	28	14	42	4
Developing	28	6	34	3
of which				
Brazil	—	2	2	—
Colombia	1	—	1	—
India	13	2	15	1
Mexico	1	1	2	—

* Hard = Anthracite and Bituminous   ** Soft = Sub-bituminous and lignite
Source: British Petroleum 1987

Table 5.19 Coal production: World, major economic groups and regions, 1938–86

	1938 HC	1938 SC	1938 Tot	1950 HC	1950 SC	1950 Tot	1960 HC	1960 SC	1960 Tot	1970 HC	1970 SC	1970 Tot	1980 HC	1980 SC	1980 Tot	1986 HC	1986 SC	1986 Tot
World 10⁶ tonnes	1207	260	1467	1435	419	1853	1966	692	2658	2160	855	3015	2732	1043	3775	3283	1220	4503
									Percentage of world									
Developed Market of which	79	83	80	73	28	63	49	23	42	48	23	41	45	26	39	41	27	37
N. America	30	2	25	36	1	28	20	1	15	26	1	15	27	1	21	24	7	19
W. Europe	43	78	49	31	21	30	23	20	22	15	19	16	10	17	12	7	18	10
Australia	1	2	1	1	2	1	1	2	1	2	3	2	3	3	3	5	3	4
S. Africa	1	—	1	2	—	1	2	—	1	3	—	2	4	—	3	6	—	4
Centrally Planned of which	17	18	17	23	72	34	47	76	54	46	75	54	48	71	55	52	71	57
China	3	—	2	3	—	2	21	—	16	18	—	13	22	—	16	25	5	20
USSR	10	7	9	13	27	16	18	27	20	20	24	21	18	19	18	18	13	17
E. Europe	5	11	6	7	45	16	7	49	18	8	52	21	9	50	20	9	53	21
Developing	4	—	3	4	—	3	5	—	4	7	—	5	7	3	6	7	1	6

HC = hard coal; SC = soft coal; Tot = total
Sources: Darmstadter *et al.* 1971; UN various years; British Petroleum 1987

Table 5.20 Coke* production: World, major economic groups and selected countries, 1950–83

	1950	1960	1970	1980	1983
World 10^6 tonnes	218	315	365	373	334
	Percentage of world				
Developed Market	79	63	61	50	45
of which					
N. America	34	19	19	13	9
of which					
Canada	1	1	1	1	1
USA	33	17	18	11	7
W. Europe	42	39	28	21	19
of which					
W. Germany	16	16	12	8	7
UK	14	10	6	3	3
Australia					
Japan	2	4	12	14	15
Centrally Planned	24	34	35	42	47
of which					
China	—	8	5	9	10
USSR	13	18	21	23	26
E. Europe	6	8	9	10	11
Developing	2	3	4	7	8
of which					
Brazil	—	—	—	1	1
India	1	2	2	3	4
S. Korea	—	—	—	1	1

* Coke from both coke and gas ovens.
Sources: UN 1976 and 1985

million tonnes increase in the 1980–85 period. After India, only South Korea, Brazil, Colombia and Mexico are significant producers among the developing countries.

Approximately one-third of the hard coal produced in the world is coking coal which is converted into coke before being charged into the blast furnaces of the iron and steel industry. The distribution of coke production has changed significantly since mid-century (Table 5.20). Steep declines in out-put in the USA and Western Europe have been accompanied by strong increases in the CPE and, up to 1983, Japan with steady expansion in those newly industrializing economies to which the iron and steel industry has been moving (e.g. India, Brazil and South Korea).

Trade

Although the proportion of world coal production that is internationally traded has remained relatively stable at 8–10 per cent since 1950, the absolute amount

Table 5.21 Trade in solid fuels*: World, major economic groups, regions and selected countries, 1938–83

	Imports						Exports					
	1938	1950	1960	1970	1980	1983	1938	1950	1960	1970	1980	1983
World 10⁶ tonnes of coal equivalent	126	115	131	201	266	267	130	123	132	202	246	241
				Percentage of world								
Developed Market of which	84	77	74	77	74	69	82	71	64	64	73	71
N. America of which	10	22	10	9	6	6	10	23	27	35	36	32
Canada	10	22	10	8	6	6	—	—	—	2	6	7
USA	—	—	—	—	—	—	10	23	27	33	30	25
W. Europe	81	54	59	43	44	37	70	46	35	19	13	10
Australia	—	—	—	—	—	—	—	—	—	9	14	19
S. Africa	—	—	—	—	—	—	—	—	—	—	9	9
Japan	6	1	6	25	24	26	—	—	—	—	—	—
Centrally Planned of which	3	17	20	19	16	16	14	28	34	35	26	28
China	—	—	—	—	—	—	—	—	—	—	2	2
USSR	1	8	4	4	2	4	—	1	10	14	10	10
E. Europe	2	9	16	15	14	12	10	27	24	21	14	16
Developing	12	7	5	4	10	15	5	2	2	1	1	1

* Solid fuels = hard coal + soft coal
Sources: Darmstadter et al. 1971; UN various years

Table 5.22 Trade in hard coal: World, 1984

Imports	10^6 tonnes		Exports	10^6 tonnes
World	305		World	305
E. Asia	120		Australia	78
W. Europe	108		USA	74
N. America	19		Poland	43
Others	58		S. Africa	38
			Canada	25
			Others	47

Source: *Oil and Energy Trends*, Vol. 10, No. 12, 1985

has increased considerably (Table 5.21). Because of its relatively high energy content and sturdy structure, hard coal dominates trade and, among the hard coals, coking coal exceeds steam coal. However, starting in the mid-seventies, the proportion of steam coal has been steadily increasing reaching almost 50 per cent by 1985.

For some 20 years after World War II the coal trade was mainly intra-regional within Europe and North America, but, by the end of the sixties inter-continental movement had become increasingly significant. Exports from the USA, Australia and South Africa expanded as markets in Western Europe and Japan developed. By the late seventies, Canada and Poland had joined the list of large exporters (Table 5.22).

The situation in Canada is unusual in that it is both a substantial importer and exporter. This is accounted for by the relative location of markets and resources. Most of the markets for both coking and steam coal are in central Canada (Ontario) while the resources are in the west (Alberta and British Columbia). Transport costs result in the major domestic markets being supplied from nearby resources in the USA while production of western coking (and some steam) coal is exported, mainly to Japan.

As the global electricity supply industry increasingly turns to coal in the face of uncertainties associated with oil price and supply and the safety and cost of nuclear technology, the demand for steam coal is expected to increase (IEA 1986). If these expectations materialize, they will serve to diminish the effect of the decline in demand for coking coal and at least sustain (if not increase) the volume of coal traded.

Uranium

The uranium fuel chain is technically the most complex of the commercial energy systems and this leads to a greater degree of interdependence between producer, processor and user than for oil or coal. Furthermore, because of its potentially hazardous nature and its use in nuclear weapons, all aspects of the uranium chain are subject not only to stringent national regulation but also to international agreements and standards. Consequently, the geography of the industry has developed within a unique set of political, institutional and technological variables. Because of these characteristics and the potential military sensitivity of production and capacity information, data are not as easily available for the uranium fuel industry as for others. Thus, most of the data in this section are exclusive of the CPE and those on processing facilities and trade are not complete. If data from the CPE were available world reserves and production would be significantly increased because of the large resource base and output in the USSR.

Knowledge of the global uranium resource base is still in an early stage of development. As a consequence, estimates are probably subject to even greater changes than those for other fuels about which there is more maturely developed information. In an attempt to improve the quality of data available, the International Atomic Energy Agency of the OECD regularly monitors reports from countries outside the CPE and publishes periodic updates (OECD 1983. In 1983, Australia had the largest estimated total resource base followed by the USA, South Africa and Canada (Table 5.23).

Production of uranium has followed an erratic path since the beginning of large-scale production after 1950 (Table 5.23). At the outset there was a decade of rapid increase as the USA stockpiled uranium (mainly for potential weapon needs). This was followed by a period of rapid decline (to a low of 15,000 tonnes in 1967) as US requirements changed dramatically (Owen 1985). By the early seventies, however, in the face of rising conventional fuel costs and uncertainty of supply, world demand for uranium was restored. By 1978, production reached the level of the previous high (1958) but, after peaking in 1980, declined again. In 1985, approximately one half of world production outside the CPE came from only five developed countries. Most of the remainder (22 per cent) is produced in three countries in western Africa (Table 5.24).

Following the initial mine production, uranium ore is milled and concentrated into uranium oxide. Further processing and fabrication involves conversion, enrichment and fuel fabrication. These stages all require advanced technology and large capital investment only available in the most industrially sophisticated economies. Table 5.25 indicates that the conversion stage is limited to North America and enrichment and fabrication to the USA, several Western European countries and the USSR.

In the early eighties, approximately one half of the uranium produced entered world trade. All African and Australian production was exported as was over half the Canadian production. The ultimate importers are those countries with nuclear-fired electricity plants but the concentrate shipped from the mines usually moves to countries for further processing (conversion and enrichment) first. Data on imports and re-exports are not available in regularly published sources and an overview of world trade in nuclear fuels can only be sketched from special studies (Neff 1984). France and Japan are the major importers, followed by the USA (largely for enrichment and re-export) and

Table 5.23 Uranium reserves: World and selected countries, 1983

| | Reasonably assured resources (RAR) 10³ tonnes U $/kg/U | | | Estimated additional resources (EAR) | | | | | | Total resources |
| | | | | Category I 10³ tonnes U $/kg/U | | | Category II 10³ tonnes U $/kg/U | | | Column (1) + column (2) |
	<80	80–130	Total (1)	<80	80–130	Total (2)	<80	80–130	Total (3)	
USA	131	276	407	30	52	82	471	339	810	489
Australia	314	22	336	369	25	394	N/A			730
S. Africa	191	122	313	99	48	147	N/A			460
Canada	176	9	185	181	48	229	179	102	281	414
Brazil	163	—	163	92	—	92	N/A			255
Niger	160	—	160	53	—	53	N/A			213
Namibia	119	16	135	30	23	53	N/A			188
France	56	11	67	27	6	33	0	12	12	100
India	32	11	43	5	14	19				62
Other	126	108	234	28	92	120	—	—	—	354
Total	1468	575	2043	914	308	1222	—	—	—	3265

RAR: Tonnages recoverable from mineable ore in deposits which are accessible to current mining and processing technology within the cost ranges shown.

EAR Category I: Tonnages recoverable from mineable ore in indicated but unmeasured deposits.

EAR Category II: Tonnages of uranium thought to be contained in situ in undiscovered, inferred deposits.

Source: OECD 1983

Table 5.24 Uranium production: World and selected countries, 1950–83

			Tonnes uranium			
	1950	*1960*	*1970*	*1980*	*1982*	*1983*
World	2,760	32,052	18,611	43,965	41,331	38,000
USA	420	14,457	9,900	16,800	10,331	7,900
Canada	200	9,786	3,530	7,150	8,080	7,500
France	100	1,038	1,250	2,634	2,859	3,200
Australia	—	934	254	1,561	4,453	3,700
S. Africa	—	4,922	3,167	6,146	5,816	5,800
Niger	—	—	—	4,100	4,259	4,200
Namibia	—	—	—	4,042	3,776	3,800
Gabon	—	—	400	1,033	970	1,042
B. Congo	2,040	915	—	—	—	
Others	—	—	110	499	787	858

Sources: Neff 1984; OECD 1983

Table 5.25 Uranium processing: selected countries, 1982

	(1) Uranium prod. 10^3 tonnes uranium per annum	(2) Conversion capacity	(3) Enrichment capacity 10^3 SWU*	(4) LWR fuel fabrication 10^3 tonnes uranium per annum
USA	10.3	21.8	27.3	3.0
Canada	8.1	5.5	—	—
France	2.9	—	10.8	0.6
W. Germany	—	—	1.3**	1.1
Belgium	—	—	—	0.4
Sweden	—	—	—	0.4
Italy	—	—	—	0.2
Spain	T	—	—	0.2
UK	—	—	1.3**	0.125
Japan	T	—	—	1.1
USSR	N/A	N/A	3.2***	N/A

* Separate Work Units
** Total of 1300 SWU capacity in W. Germany, UK and Netherlands
*** Export capacity only
Source: Neff 1984

Western European countries with large nuclear electric capacity such as Western Germany and the UK.

Electricity

Unlike the commercial primary energy sources reviewed so far, electricity is mainly a secondary form of energy. However, the scale of the industry and the fact that parts of it are conventionally referred to as primary energy, warrant the inclusion of the industry in this chapter. For the sake of completeness, the whole industry is briefly reviewed distinguishing between primary (hydro and geothermal), fossil-fuel and nuclear generation systems. It should be noted that some major agencies (e.g. United Nations) still consider electricity generated by nuclear systems as primary energy.

Production

Since 1950, the production of electricity has increased without interruption. Remarkable versatility, cleanliness and convenience in all sectors (except private transportation) make energy in the form of electricity highly attractive. At the world scale over 90 per cent of uranium production, more than 50 per cent of coal and a significant proportion of the production of oil and natural gas is converted into electricity. Production is concentrated in the DME and, to a lesser extent, in the USSR and Eastern Europe (Table 5.26). Although still only producing less than 15 per cent of world electricity, the rate of increase in the DE in the eighties has been markedly higher than that of the other two major economic groups. The most rapid expansion has taken place in the newly industrializing countries of Latin America, South and East Asia.

In broad terms, the most notable feature of the mix of production systems used to produce electricity since 1950 is the persistence of the role of fossil fuel-fired plants which have produced approximately two-thirds of the world's electricity for the last 50 years (Table 5.27). However, while coal is still the major fuel used, it no longer has the dominance it had in mid-century as oil products and natural gas have become the preferred fuels in many countries. Despite large absolute increases in production, the relative importance of hydro has declined as the proportion supplied by nuclear systems has increased.

In individual countries, the mix of production systems may vary considerably from the global aggregates. Each country varies in respect of resource opportunities, access to technological developments, comparative cost and public policy. Thus the hydro dominated systems of, for example, Brazil, Canada and Sweden contrast with the fossil fuel-nuclear systems of the UK and South Korea or the nuclear dominant system of France.

Because of the impossibility of long-distance marine transportation, international trade in electricity is largely limited to intra-regional, land transfers. Although there has been a considerable increase in the range over which

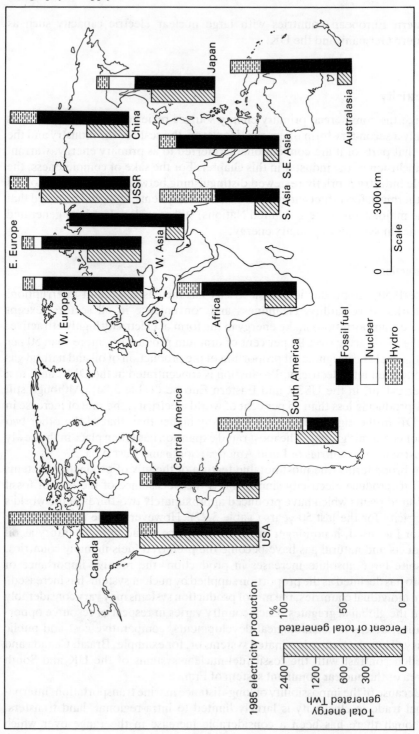

Figure 5.10 Electricity production: World, 1985
Source: UN 1987

Table 5.26 Electricity production: major economic groups, 1950–85

	1950	1960	1970	1980	1985
World 10^9 kWh	959	2300	4908	8247	9675
			Percentage of world		
Developed Market	81	74	71	65	61
of which					
N. America	46	43	36	34	31
W. Europe	27	25	23	21	20
Oceania	1	1	1	2	2
Japan	5	5	7	7	7
S. Africa	1	1	1	1	1
Centrally Planned	15	21	22	25	26
of which					
China	—	1	2	4	4
USSR	10	13	15	16	16
E. Europe	6	5	5	5	4
Developing	5	6	7	10	12
of which					
Brazil	1	1	1	2	2
India	1	1	1	1	2
Mexico	—	—	1	1	1
S. Korea	—	—	—	1	1

Sources: UN various years

Table 5.27 Electricity production by primary energy source: World and selected countries, 1950 and 1985

	1950				1985			
	Total 10^9 kWh	Hydro	Fossil fuel	Nuclear	Total 10^9 kWh	Hydro	Fossil fuel	Nuclear
			Percentage				*Percentage*	
World	959	36	64	—	9675	21	64	15
USA	390	26	74	—	2525	11	73	16
Canada	55	96	4	—	460	66	21	13
Japan	45	16	84	—	673	13	63	24
France	33	48	52	—	326	19	16	65
UK	66	2	98	—	295	1	78	21
Sweden	18	94	6	—	137	52	5	43
USSR	91	13	87	—	1544	13	76	11
Brazil	8	88	12	—	193	92	8	—
India	7	71	29	—	188	31	67	2
S. Korea	1	23	77	—	63	6	67	27
Venezuela	1	—	100	—	45	44	56	—

Sources: UN various years

143

electricity may be transported, still less than 3 per cent of total production enters international trade. More than half of this takes place between the elaborately interconnected national systems in Western Europe and most of the remainder consists of increasingly large exports from Canada to the USA and from the USSR to Eastern Europe.

Summary

This chapter has reviewed the evolution of the broad spatial structure of commercial energy supply systems in the years since 1945. During this period, total global primary energy production increased approximately four-fold. A new source of commercial energy emerged with the uranium industry and the oil, gas and coal industries expanded world-wide so that, by volume and value, energy commodities dominate world trade. Despite the rapid growth of production, reserves of all the fuels have also grown although by the mid-1980s reserve additions of conventional oil were only barely staying ahead of production.

In 1986 the majority (80 per cent) of *conventional oil reserves* were located in the Developing Economies and over one-half in the Persian Gulf area. *Reserves of unconventional oil* sources (oil shale and tar sands), on the other hand, were concentrated in Canada, the United States, and Venezuela. *Oil production* was concentrated in the Developing Economies (45 per cent in 1986) with almost one-third coming from members of OPEC and 19 per cent from the Persian Gulf area. The dominance of OPEC producers declined somewhat in the early 1980s as production from other areas increased (e.g. Mexico and the United Kingdom) but in the second half of the decade it appears that this trend will be short-lived. Members of OPEC and the USSR were the greatest *exporters* of oil, both depending heavily upon the foreign exchange generated by the trade to fund their economic development programs. Japan, Western Europe and, to a growing extent, the United States were the major *importers* with the non-oil-producing Developing Economies as an important second group.

Natural gas reserves are more concentrated than those of oil with 57 per cent in two countries, the USSR and Iran. Significant amounts still exist in North America, the North Sea, South-east Asia and Australia. *Market production* of gas was focused in North America (42 per cent), the USSR (29 per cent) and those countries sharing the North Sea reserves (13 per cent). Being less easily transferable across water than the other fuels, only 15 per cent of natural gas production was *traded* in three major systems. One focused on supplying Japan with LNG by tanker from around the Pacific Ocean and, to a lesser extent from the Persian Gulf; the second centred on Western Europe and supply pipelines from the USSR and the North Sea and both pipe and tanker from North Africa; the third was the North American system consisting mainly of Canadian pipeline exports to the United States.

Plate 1   Open pit coal mine in North-east British Columbia, Canada. Washing, sorting and loading facilities — lower left.
Source: Teck Mining Group, Vancouver, B.C., Canada

Plate 2   Semi-submersible drilling rig in South China Sea.
Source: The British Petroleum Company, London

Plate 3   Flaring crude oil during drill stem test from floating ice-breaking drilling
unit anchored in the Beaufort Sea, N.W.T., Canada.
Source: Gulf Canada Resources Ltd., Calgary, Alberta, Canada

Plate 4   Oil sands mine in Northern Alberta, Canada.
        Source: Syncrude Canada, Fort McMurray, Alberta, Canada

Plate 5   Oil sands processing plant in Northern Alberta, Canada.
        Source: Syncrude Canada, Fort McMurray, Alberta, Canada

Plate 6   Bitumen production pads, Cold Lake, Alberta, Canada.
Source: Esso Resources, Calgary, Alberta, Canada

Plate 7 Steam production, storage tanks and — in lower right — production wells at Cold Lake, Alberta, Canada
Source: Esso Resources, Calgary, Alberta, Canada

Plate 8 Line of pumpjacks pumping steam-readied bitumen to surface at Cold Lake, Alberta, Canada.
Source: Esso Resources, Calgary, Alberta, Canada

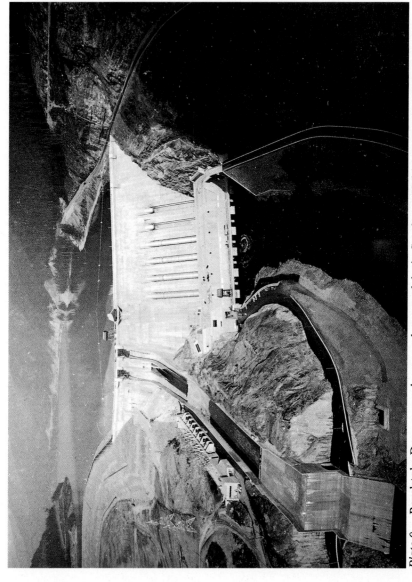

Plate 9   Revelstoke Dam, power house and storage lake in South-east British Columbia, Canada.
Source: B.C. Hydro and Power Authority, Vancouver, B.C., Canada

Plate 10   Nuclear electricity generating station on the shore of Lake Ontario, Canada.
Source: Ontario Hydro, Toronto, Canada

Plate 11 Deep sea coal loading port, Point Roberts, British Columbia, Canada.
Source: Westshore Terminals, Delta, B.C., Canada

Plate 12 Liquified natural gas carrier.
Source: The British Petroleum Company, London

Coal reserves, the largest of the fossil fuel reserves in energy terms, were equally shared between the Developed Market Economies and the Centrally Planned Economies (48 per cent each) with very little in the Developing Economies. The USA and USSR each had approximately one-quarter and China, 17 per cent. *Coal production* was roughly equally divided between Europe, North America, USSR and China. Western European production was only 10 per cent of the world total in 1986 (almost 50 per cent in 1938) and the trend was downward. Almost 10 per cent of coal production was *traded* in 1986, exports having extended from an essentially intra-continental scale to a trans-oceanic system in the preceding 25 years. As Western Europe and Japan have increased their imports, so the United States, Australia, South Africa and Canada have developed large export based coal industries.

Because of the military significance of uranium, data for *uranium reserves* do not include the Centrally Planned Economies. Outside these countries reserves are concentrated in North America and Africa (each with 30 per cent) with most of the remainder in Australia and Brazil. *Uranium production*, which peaked in 1980, was distributed in the same way as the reserves. More than half the uranium produced was *traded* internationally with countries such as Australia, Namibia and Niger exporting all their production to the major consuming countries of Western Europe and Japan. The United States was also a major trader importing uranium and exporting enriched reactor fuel.

The *production of electricity*, a secondary form of energy derived for the most part from fossil and nuclear fuels, has increased more rapidly since 1945 than any other sector of the energy industry. However, except in some of the newly industrializing countries among the Developing Economies, rates of increase slowed down considerably after 1980. The major *primary energy sources* used by the electricity industry remain the fossil fuels (approximately 65 per cent over the whole period) followed by hydro (down from 36 per cent to little more than 20 per cent) and, finally, the newcomer, nuclear fuels (13 per cent). Other renewable primary sources such as geothermal, wind or solar have yet to become globally significant. Two-thirds of *world production* takes place in the Developed Market Economies, one-quarter in the Centrally Planned Economies and only 10 per cent in the Developing Economies. Only approximately 5 per cent of the electricity produced is *traded internationally* and that on an intra-continental basis largely within Western Europe and between Canada and the United States.

CHAPTER 6

Energy futures

Over time the character of energy systems changes, rapidly as new delivery systems come into operation or large price adjustments occur or slowly but persistently as one type of energy-producing or using equipment is replaced by another. Because energy is such a vital component of the economic and social fabric of society, governments must explicitly consider the future state of energy systems, identify their options and determine what action to take in order to achieve their policy objectives. Similarly, firms in the energy supply industries have to estimate the future energy resource base and the demand for the energy they market. Both the public and private sectors must develop some perceptions of the future state of energy systems if they are to exercise prudent and effective management.

Projections

Projecting energy futures was for a long time the domain of the planning divisions of the major energy supply firms but has now extended to regional and national governments, international agencies and independent research institutions. Statements by these organizations about energy futures are in the form of conditional probabilities. They do not state what *will* happen but, rather, what *could* happen if specified conditions are met. All too often readers (and, sometimes, authors) of reports on energy futures lose sight of this distinction. Some of the misinterpretation arises from the use of terms in an unspecified way. Generally, however, 'prediction' and, to a lesser extent, 'forecast', are used to imply probability; 'projection' connotes possibility; and 'scenario', feasibility (Eden *et al*. 1982). Whatever the form, statements about future supply are intended to identify the direction as well as the relative and absolute magnitude of change.

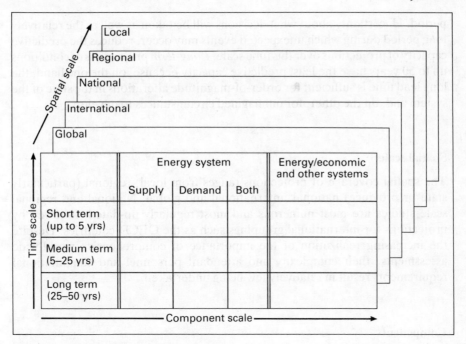

Figure 6.1 The scope of energy projections

The scope of projections

Projections vary in terms of temporal, spatial and phenomenal scope (Fig. 6.1). It is helpful to consider briefly each of these three dimensions before reviewing the methodology used in preparing the projections.

Time-scale

Projections may be applied to time periods as short as days or weeks and as long as half a century. Within this range a distinction may be made between short, medium and long-term projections representing up to 5, 25 and 50 years respectively (UK, DOE 1978).

Short-term projections, used for operational and tactical purposes, have the highest probability of being correct. Despite the predictive expectations associated with short-term projections, unforeseeable events such as political and labour disruptions, abrupt price changes or extreme weather conditions may render them unreliable. *Medium-term* projections of conditions up to 25 years deal with a time-frame long enough for existing strategies to be implemented and new ones formulated. Thus, although rooted in the technologies, supply patterns and use habits of the date of formulation, significant changes can arise from decisions taken early in the projection

147

period. Uncertainty about what decisions will be taken as well as the relatively long period during which unexpected events may occur, reduces the predictive capacity of projections over this time scale. *Long-term* projections of conditions up to 50 years have the least predictive capacity because, on the one hand, the long lead time is sufficient for order-of-magnitude alterations in the state of the system and, on the other, for unimagined circumstances to develop.

Spatial scale

The spatial coverage of projections ranges from local, regional (particularly state or province) national, international and global. National and regional scale studies are most numerous and most regularly up-dated, followed by projections for international groupings such as the OECD countries. Despite the increasing realization of the importance of comprehensive world-wide assessments, their complexity and attendant personnel and computational requirements result in relatively few being undertaken.

Components

The phenomena included in energy projections may be divided into two major categories: the components of the energy system itself and the interaction of that system with the economic and other societal systems. A further sub-division within both categories is between projections focused on supply and those centred on demand. On a finer scale, the energy system category may be approached in three ways (Hoffman and Wood 1976):

(1) by sector – dealing with the energy requirements and supply sources of each consuming sector and sub-sector (e.g. transportation: aviation).
(2) by total energy – dealing with the demand for and supply of total energy (e.g. total energy requirements of Canada in 1990 in 10^9 tonnes oil equivalent).
(3) by energy source – dealing with the demand for and supply of individual energy sources and forms (e.g. coal or electricity).

The second broad category of phenomena centres on the relationship between the energy and economic systems and addresses such fundamental interactive issues as the impact of energy upon economic activity and vice versa.

Methodology

The study of energy futures is mainly carried out by means of models which consist of a simplification of reality in order to present significant relationships in a generalized form. In most cases the relationships are expressed by equations, usually solved by means of computer programs. Four characteristics of energy future modelling are important. First, modelling involves simplification and, therefore, a selection from the total complex of parameters and variables that might be included. Second, if the models are to produce useful results, the fundamental relationships between the parameters and variables must be well developed and verified. Third, and affecting the variables, there is considerable interdependency between supply and demand conditions. Finally, and arising from these characteristics, future energy modelling is fraught with uncertainty. Despite the sophisticated methodologies applied, many modellers remain sceptical of their own work and some informed commentators raise fundamental questions about the limitations of energy modelling (Bohi and Toman 1983; Weinberg 1980).

The majority of predictive modelling is based upon geophysical, thermo-dynamic or economic theory. Geophysical models are used to assess the potential geological resources ultimately available (Levine 1977). Physical process and thermodynamic models are suitable for projecting the potential for production of energy sources or the performance of energy-converting equip-ment (UK, Department of Energy 1978). Economic models on the other hand, lead to predictions concerning, for example, the amount of fuel that would be used for a particular purpose in relation to its price or the relation between GNP and overall energy demand (Bohi and Zimmerman 1984). It is increasingly common for the approaches to be combined or used in conjunction with one another for individual parts of comprehensive models.

Projective models may be classified on the basis of their scope (Fig. 6.1) or the methodology used. In terms of the latter they may be roughly grouped into mathematical programming (e.g. linear and non-linear), activity analysis (e.g. input–output) and econometric analysis using regression and other statistical techniques (Beaujean and Charpentier 1978). The complexity of many models has made it necessary to develop another level of expertise, that of the model interpreter. Such an individual (or group) interprets the implications of projec-tive models to the decision makers responsible for managing energy systems (Greenberger 1977; Roberts and Waterman 1980). In order to reduce the range of projections provided by different models consensus-seeking mechanisms are increasingly used. One such mechanism, the Delphi technique, aims to achieve consensus of informed opinion through successive questioning of individuals (Manne and Schrattenholzer 1984).

Future supply

The most fundamental question to be answered about future energy supply is how long will the various sources last and at what cost? Projected answers depend upon a number of interdependent variables including the stock or flow of each energy source, the proportion of the stock which can be extracted (the recovery rate), and the rate of production.

Concepts and terminology

Projections of the stocks and flows of energy potentially available call upon the skills of many different disciplines. Partly because of this, resource estimates often differ in their use of concepts and terminology and this leads to difficulties when comparing results. Consequently, to help to maintain a conceptual consistency, it is useful to have a general framework into which the terms of any specific projection may be fitted (Blunden 1985).

The framework presented in Figure 6.2 uses information and availability as two major dimensions in terms of which two major classes are identified: undiscovered potential geological resources and discovered reserves. Each major class is further subdivided: the first into speculative, hypothetical and inferred and the second into conditional and current. These distinctions are most frequently applied to stock resources (with which this section is mainly concerned) but are, nevertheless, also relevant to flow resources. For the

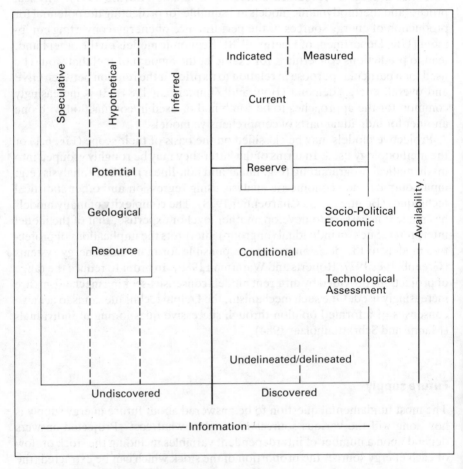

Figure 6.2 Classification of future energy sources

latter, unmeasured and measured are analogous to undiscovered and dis-covered and the concepts of conditional and current resource are as useful in considering flow as stock resources.

Potential resources consist of undiscovered sources thought to exist on the basis of geo-scientific theory and data. As information increases, uncertainty decreases as suggested by the three categories speculative, hypothetical and inferred. The chief means of increasing information is exploration and the proportion of the potential resource base that is discovered is directly related to the exploration effort expended.

Once discovered a deposit becomes a *conditional* reserve, that is one which may enter the energy supply system providing a number of conditions are met. Before examining the major conditions more thoroughly it is useful to sum-marize them. The first condition involves the gathering of still more information about the size and quality of the discovery (delineation and assessment). The larger the size and the higher the quality the more likely a discovery will eventually result in a current reserve. The second condition to be met is technological: what proportion of the delineated deposit can be produced (the recovery rate)? This is a function of the physical characteristics of the deposit and the performance of available technology.

The next test is economic: are the costs of production such that a satisfactory return can be realized within the expected price structure? What constitutes a 'satisfactory return' will vary according to the expectations of the producer. For example, publicly owned organizations may be satisfied with a lower (or even negative) direct return if other, associated, benefits are achieved (e.g. employ-ment, improved balance of trade).

After the economics of production for an energy source are seen to be satisfactory there are still other conditions to be met before the production decision is taken. These may be grouped under the headings of political and social conditions. The role of political versus 'free-market' forces in condition-ing the mobilization of energy resources varies with time and place. During the 1973-83 decade all sections of the energy industry became increasingly politicized and political intervention grew at all scales. Local and state govern-ments determined whether an energy source could be produced and under what conditions; national governments applied taxes and royalties, determined the rate of production and whether exports (or imports) would be permitted. By the mid-1980s this tendency was reversed and, as their title would suggest, the governments of the Developed Market Economies considerably reduced their interventionist role in favour of the operation of 'free-market' forces. Nevertheless, if the operation of these forces leads to developments which are not in keeping with public policy objectives, politically inspired conditions will be re-imposed. Finally, and in many countries, at the root of politically imposed conditions, social concerns about such matters as environmental quality and risks to health and life may result in such stringent regulatory controls that production costs are raised above the competitive threshold.

When all these conditions are satisfied and production commences an energy

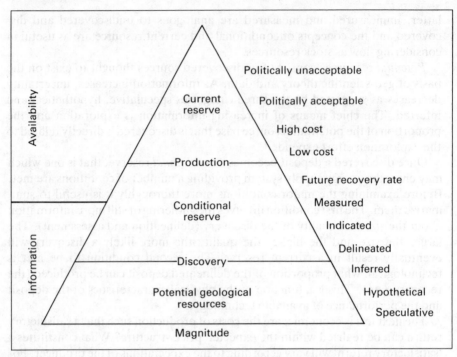

Figure 6.3 Conceptual relation of the magnitude of estimates of future energy supply categories

source becomes a *current* reserve within which a distinction is often made between indicated and measured, again reflecting improved information. An important aspect of this progression from conditional to current status is that if circumstances change, it is reversible and a current reserve may revert to being a conditional one. For example, as the price for natural gas in the USA fell in 1984–85 much Canadian gas was unable to compete in the North American export market and thus effectively moved from being a current to a conditional reserve.

Positioning a particular projection in a potential-conditional-current framework assists in the initial assessment of its reliability or, if comparing several projections, serves as a useful check to ensure that they are, indeed, comparable. A major reason for caution when interpreting future supply projections is the great difference in magnitude between estimates in each category. Generally, the greatest magnitude is associated with the potential geological resource class and the least with the current reserve category (Fig. 6.3).

Variables

Beyond estimates of the ultimate magnitude of the potential geological resource, the most important variables about which assumptions have to be

152

made in the preparation of projections of future energy supply are exploration effort, technological development, economic conditions, and socio-political objectives (including concerns about environmental impact). In this section each is considered as background to the review of some recent global projections which follows.

Exploration effort Exploration which leads to discovery increases information about the energy resource base most positively because it confirms the existence of a portion of the potential geological resource. However, even if exploration does not lead to discovery, it can add valuable information which permits the refinement of projections about the undiscovered. In preparing projections of the future energy resource base two exploration related issues have to be addressed. First, what proportion of the large deposits remain to be discovered and, second, are the exploration programs (now or in the future) located in areas most likely to contain them? One school of thought tends to assume that the answer to the second question is obvious because companies will, in their own interest, explore where the chance of success is greatest (Broadman 1985). There are others who note that when the factors of cost and returns are considered this may not be the case (Odell and Rosing 1983). Some evidence in support of this latter view is provided by the incentives which the Canadian government felt it had to provide in order to achieve a politically acceptable oil-exploration program in the Mackenzie Basin/Beaufort Sea area of the Arctic. Reflecting different conventions among the earth scientists involved, it should be noted that projections of the future supply of conventional oil and gas usually include estimates of the yet-to-be-discovered resources, whereas those for coal, uranium, oil sands and shales do not.

Technological development Geographers are accustomed to viewing technology as the development and production of tools which are used by humans in their interaction with the environment (Hare and Hewitt 1973). In the context of energy futures, the important considerations are the role technological development may play in augmenting and changing energy supply and in altering energy use patterns and requirements. The latter determines not only the general rate of use of energy but the demand for individual sources. The role of technological development is a complex matter because technological change itself results from conscious decisions to select potential technologies and invest time and money to develop them. Consequently, the projection of the impact of technology upon future energy supply is not only a matter of projecting the effectiveness of particular developments but also of assessing the likelihood that society will make the necessary investment to achieve them, the speed with which they might be adopted, and the time required to have an impact on future supply. For example, the extent to which it is realistic to include bitumen and kerogen in medium-term estimates of future oil supply depends first on the technological feasibility of extraction and second, on the cost competitiveness of the product. In the case of nuclear fuels, the demand

153

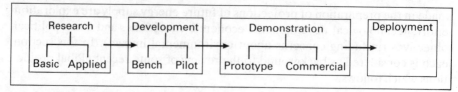

Figure 6.4 Dynamics of technological development

for uranium is closely tied to the efficiency of the nuclear reactors in use and whether breeder reactors are introduced (Weinberg 1986).

The evolution of new energy technology may best be described in four stages: research, development, demonstration and deployment (Fig. 6.4). The more unconventional the technology the longer the time required to progress through the stages and the greater the costs involved.

The *research* stage consists of two parts, basic and applied. The former refers to the development of the underlying theory and its general validation which is usually carried on in universities and other large-scale research institutions. Applied research focuses upon the application of the theory, the design of the necessary equipment and the building of an experimental facility.

The *development* or engineering stage involves several intermediate steps in which progressively larger test units are built to evaluate different designs and component performance and to develop scaling laws. In the *demonstration* stage larger-scale prototype units are built, embodying the most promising designs complete with ancillary equipment. Still larger commercial-scale plants are then built to assess reliability and durability of the whole system and to demonstrate its economic characteristics as realistically as possible.

Deployment refers to the adoption of the new technology by industry. The adoption rate will be a function of the technical qualities of the technology, its ability to compete economically with existing systems, and its socio-political acceptance. Market penetration by a new technology takes place in two ways: first by being chosen for new, additional capacity to a system which is growing and second, by replacing equipment during modernization of an existing system. The general form of the deployment of a new technology over time is represented by the logistic substitution curve (Fig. 6.5). This S-shaped curve has been shown to fit many actual cases of technological innovation and, with modifications, is used in the projection of future supplies (Häfele 1981).

Figure 6.6 illustrates the application of these concepts to the most recent major innovation in energy supply systems, nuclear power. Thirty-seven years elapsed before Einstein's theory was validated by the first demonstration of nuclear fission, a further 20 years before the first commercial scale reactor was ordered for full-scale utility use and, after 20 more years, market penetration has only reached approximately 10 per cent of the global market. In other words it has taken a human lifetime for one major new energy technology to evolve and secure 10 per cent market penetration. Three-quarters of a century elapsed despite wartime urgency and minimal funding constraints or regulatory

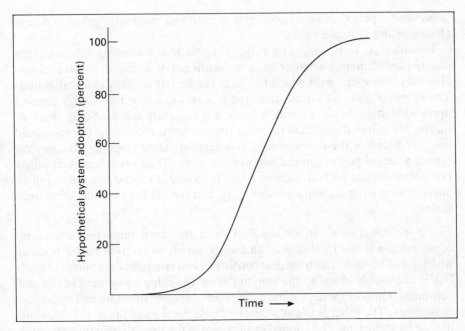

Figure 6.5 The logistic substitution curve

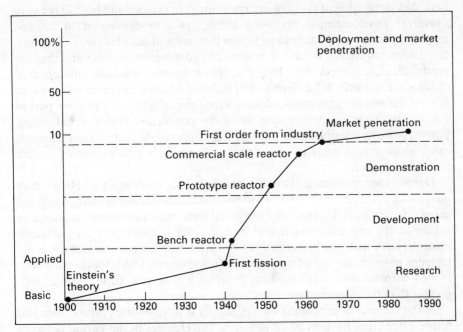

Figure 6.6 The evolution of nuclear power
Source: Rudman and Whipple 1980

delays and a period of rapid economic growth and minimal regulatory delays (Rudman and Whipple 1980).

In summary, technology is a major variable in determining future energy supply (and demand) whether from conventional or unconventional sources. However, the speed with which its effect can be felt is easily over-estimated. The length of time and funds required to develop a new technology depend upon what stage in the evolution process it currently stands. Some, such as fusion, are still in the applied research stage; others, such as the photo-voltaic cell, are between the development and demonstration stage and tertiary oil recovery techniques are in the deployment stage. Thus, what has been called the 'technological fix' (Weinberg 1966) is indeed a crucial consideration in energy futures but its impact only emerges in the medium to long-term time scale.

Economic conditions In its simplest form the most important economic consideration is the total cost of an energy supply in relation to the price at which it can be sold. The total cost consists of two components applied to each link in the supply chain all the way to the energy-using equipment of the end consumer. One component is the actual cost of the construction and operation of facilities. The other is the administered cost made up of taxes, royalties and licence fees which may be applied at any point in the supply chain thus affecting the price to the next link.

Actual costs are influenced by factors reviewed in Chapter 4. Suffice it to say here that development costs of energy sources in remote and hostile (or very sensitive) environments, involving either yet-to-be-developed or untried technologies, are very likely to be higher than costs in familiar areas. Administered costs can change rapidly in response to government dictates and thus are most difficult to project in the long run. However, once imposed, administered charges are unlikely to be significantly reduced because governments come to rely on the income generated, although they may shift them from one part of the supply chain to another (e.g. from the production end to retail sales). Furthermore by administering subsidies (the opposite of charges) governments can reduce actual costs in order to encourage new production or sustain prevailing output.

Despite the importance of total costs, what really matters is the relation they bear to the price received. In order to be economically feasible in a commercial sense it is necessary for the price to exceed total costs and permit an adequate return to the resource owner and investor. What constitutes an 'adequate return' varies with the organizations involved but, for governments, usually includes payment for use of the resource (resource rent) and, for investor-controlled corporations, a price which provides a return on capital equal to or greater than the discount rate.

In the context of projecting energy supply it is, of course, *future* costs and future prices that have to be considered. The changes in the period between 1973 and 1986 illustrate the volatility of the structure of energy prices and emphasize the difficulty of estimating the future (Eden *et al.* 1982). One strategy that can be adopted for projections is to estimate the amount of a conditional energy source that would be available in selected price ranges (e.g. Fig. 5.23).

Socio-political objectives The socio-political values and objectives of the populace and governments constitute another set of variables which influence future energy supply. In Western democracies, the concerns and preferences of the population-at-large (often aired and magnified by the media) become reflected in government policy. In other political systems public opinion is little developed or, if it is, plays only a small part in shaping government actions. In either case, individual national governments (or consortia of governments) set the context within which many decisions affecting energy futures are taken.

Of particular importance to future supply considerations are government decisions about energy production rates and conservation strategies. The production rates allowed by the governments of the major energy-producing countries will be an important determinant of the amount of, say, oil or coal that is made available to the world market. In the long run, these rates will also play an important part in determining how rapidly resource stocks will become exhausted (reserve/production ratio). The actual amounts produced and marketed will, however, depend upon the demand. Consequently, if the governments of the major consuming countries develop strategies to encourage energy efficiency and conservation, supply requirements will be reduced. It is often, and correctly, said that the most effective way of prolonging energy supplies is to reduce the demand for them.

The operation of these general influences may be illustrated by considering the example of oil. The timing of the eventual commercial exhaustion of conventional oil will be determined by the extent to which (1) the major producer countries regulate the rate of production and manipulate prices and (2) the major consumer countries determinedly encourage off-oil and oil efficiency programs and the development of alternative energy sources. If demand is dampened and prices generally maintained, production rates will decline and the life of the oil stocks will be extended.

In addition to the formally adopted policies of national governments aimed directly at energy, future energy supply systems will be influenced by the concerns and preferences of the population at large and policies and regulations developed by governments in response to them. In many countries advocacy groups organized around such issues as environmental quality, risks to health and life, 'hard' and 'soft' energy and 'centralized' and 'dispersed' systems actively intervene in the evolution of supply patterns at least at the local and regional levels (Okrent 1980; Lovins 1978). Among these issues, concern for environmental quality and the associated risks to health and life are of increasing importance and now reach beyond the local and regional to hemispheric, if not global, scales (Edmonds and Reilly 1985).

To the extent that public awareness and scientific verification of the environmental effects of the links in the production-consumption chain of the various energy sources are translated into regulatory controls by governments, they will influence future energy supply. Three environmental issues are of particular concern because of their increasing scale and their complexity: the growing carbon dioxide content of the atmosphere, acid rain, and the disposal of nuclear waste (Gates 1985).

The carbon dioxide content of the atmosphere has been steadily increasing, to a large extent as the result of the combustion of fossil fuels. The effect of the increasing concentration is to bring about a rise in global temperature with attendant implications for other atmospheric components. Acid rain refers to precipitation with a high acidity resulting mainly from the sulphur and nitrogen oxides emitted from the burning of fossil fuels. The major concern about acid rain is the cumulative ecosystem damage which it may cause, particularly to lakes and forests. Under appropriate atmospheric conditions, the oxides may be transported over semi-continental distances.

Carbon dioxide concentration and acid rain have much in common. The primary impact of both is on the atmosphere, they both result from the combustion of fossil fuels, and their effects extend over a wide area. Furthermore, scientific understanding of the processes and impacts is still limited and the definitive establishment of emission sources still open to dispute. As a consequence it is difficult to raise public awareness to a level sufficient to invoke action. Even if this is achieved at the national level, effective solutions are difficult to formulate because of the scientific uncertainty which still prevails and the difficulty of achieving the necessary international cooperation. Taken together, these two impacts may significantly delay or even prevent a large-scale increase in the use of coal in future energy supply.

The expanding caches of untreated nuclear wastes constitute a potential hazard which causes general public concern and forms part of the basis of the rising social resistance to nuclear energy. Despite the confidence expressed by supporters of nuclear energy for reprocessing and ultimate geologic storage, social acceptance of the risks hangs in the balance, a balance which could be tipped by one or more widely publicized nuclear accidents. Were this to happen, the market penetration of nuclear energy would continue to progress slowly.

Methodology

Projection of future energy supplies involves estimating the answers to three basic questions: (1) how much exists? (2) how much can be produced? and (3) how much will be produced? The answers to these questions are to some extent interdependent making it desirable that they should not be considered in isolation. However, all too frequently, disciplinary solitudes and the simplification required in model formulation prevent this.

The long-term question of how much of an energy source exists is usually addressed by geological scientists using a volumetric approach, the inter-mediate-term production question by applied scientists using engineering approaches and the shorter-term question of how much will be produced, by economists using econometric techniques. All three make use of past experi-ence (and, to some extent, therefore, do not provide for unexperienced events) and, despite the sophistication of some of the methodology, require substantial assumptions and involve a considerable degree of individual judgement.

Geological scientists estimate the volume of sedimentary rock thought capable of containing energy fuels and, from previous experience with such rocks, estimate the amount of the target energy they contain. Variations among individual projections are often considerable and result from conceptual and terminological differences as well as differences in judgement (Odell and Rosing 1985; Gold 1985). Projections based upon volumetric considerations may be refined by adding a probabilistic appraisal of the potential resource in each geological province or field (USGS 1975; Canada, EMR 1977).

Applied science approaches recognize that the production of stock energy sources must ultimately decline. Further, the timing of the onset of the decline is a function of several linked variables including past production, production rate, exploration effort and discovery rate. The ultimate potential geological resource is not usually explicitly considered in such projections and they have only recently considered the influence of price upon the production variables. One of the most widely used forms of projection in this category fits the logistic growth curve to the production history of a finite resource (Hubbert 1962).

By plotting the production histories of a number of finite energy sources a standardized production curve such as that shown in Figure 6.7(a) is identified in which the area under the curve represents the ultimate cumulative produc-tion. When the total reserves, past production and recovery rates are known the form of the remainder of the curve may be extrapolated for different future production rates, thus predicting the time of maximum production. The general forms of the curves for cumulative discovery and production and proved reserves are illustrated in Figure 6.7(b) and for the rates of change in these components in Figure 6.7(c). The latter indicates that the peak of proved reserves occurs when the curves of the rate of discovery and rate of production cross. The timing of this peak is seen to be approximately half-way between the peaks in the discovery and production rates.

Projections using this approach underline the unavoidable – the ultimate exhaustion of stock resources – but, depending upon the assumptions about exploration effort, discovery rate, technological innovation and price struc-tures, can produce either optimistic or pessimistic estimates. Furthermore, fitting selected curves to past experience involves a greater degree of subjec-tivity than the mathematical precision of curve-fitting and the quantitative historical data-base would suggest.

Econometric approaches usually relate potential supply to exploratory and development investment in the context of different price structures. Such approaches often pay little attention to ultimate resource base estimates or the

159

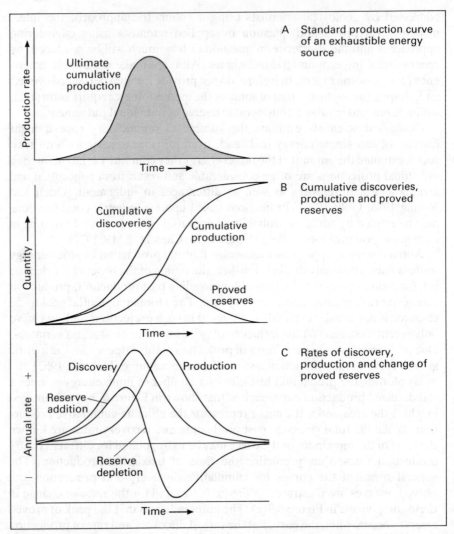

A Standard production curve of an exhaustible energy source

Ultimate cumulative production

Production rate →

Time →

B Cumulative discoveries, production and proved reserves

Quantity →

Cumulative discoveries

Cumulative production

Proved reserves

Time →

C Rates of discovery, production and change of proved reserves

Annual rate

Discovery

Production

Reserve addition

+

−

Reserve depletion

Time →

Figure 6.7 The general form of energy supply projections based upon the application of the logistic growth curve to historical discovery and production data
Source: Hubbert 1962

physical-technical aspects of producibility. Rather, they rely heavily upon historically determined relationships between price and supply.

Projections

Between 1975 and 1985, numerous global projections were published (e.g. WAES 1977; WEC 1978; Wilson 1980; Häfele 1981; IEA 1982; WEC 1983a and b; Edmonds and Reilly 1985). It is beyond the scope of this section to attempt a critical review of this large body of technical material. Instead, a consensus view will be presented, relying mainly on generalizations derived from *World Energy Outlook* (IEA 1982).

Views of the global resource base of non-renewable fuels in the early 1980s are contained in Table 6.1. The fuels are listed in ascending order of estimated lifespan (reserve/production rates) which emphasizes that conventional oil, currently supplying over 40 per cent of world commercial energy requirements, has the shortest life expectancy. In addition Table 6.1 contains estimates of the resource base of the fuels which may be used to replace conventional oil. Unconventional oil sources (shale and sand) constitute large and relatively untouched stocks, and coal, as the raw material for synthetic oil, has the largest resource base of all the fossil fuels.

Oil Estimates of the remaining total conventional oil resource (i.e. reserves plus potential geological resources) range from a low of 1700×10^9 barrels (approx. 230×10^9 tonnes) to a high of 5600×10^9 barrels (approx. 765×10^9 tonnes). However, if the extreme estimates are excluded a consensus seems to develop around the figure of 2000×10^9 barrels (approx. 275×10^9 tonnes) (Edmonds and Reilly 1985). The highest projections represent the most optimistic view of the rate of increase of recovery rates (from the current 30 per cent to 60 per cent and over) and about the discovery potential in the still largely unexplored areas of Africa, Latin America and South-east Asia. The more moderate and lower projections are less optimistic about the potential of enhanced recovery techniques and believe that few if any giant and super-giant fields remain to be discovered (IEA 1982).

Uncertainty will only be reduced by exploratory drilling in the currently unexplored areas. However, during the first half of the 1980s, the total number

Table 6.1 Projected global reserves, reserves/production ratio and potential recoverable geological resources of mineral fuels

Fuel	Reserves (10⁹ tonnes oil equivalent)	Reserves/production ratio (years)	Potential geological resources (10⁹ tonnes oil equivalent)
Conventional oil	95	33	165
Uranium*	60	45	41
Conventional natural gas	102	60	303
Coal	532	226	709
Oil sand	40	71,000	70
Oil shale	46	(no significant production)	280

Table header units: Reserves $(10^9$ tonnes oil equivalent$)$; Potential geological resources $(10^9$ tonnes oil equivalent$)$

* Excluding Centrally Planned Economies
Sources: IEA 1982; WEC 1983a, 1983b; British Petroleum 1987

Table 6.2 Oil – regional distribution of recoverable conventional reserves and potential geological resources

	Reserves	Potential geological resources
	(Percentage of world)	
Middle East	56	21
Centrally Planned Economies	11	32
Latin America	13	10
Africa	8	3
North America	6	14
Asia-Pacific	3	12
Western Europe	3	8
Total World (10^9 metric tonnes)	89	178

Source: IEA 1982; British Petroleum 1987

Table 6.3 Oil – conventional and unconventional reserves: selected countries

	Conventional oil		Unconventional oil	
	Reserves	Reserve production* ratio	Sands	Shale
	(10^9 tonnes)	(years)	(10^9 tonnes oil equivalent)	
Saudi Arabia	22.7	90	—	—
Kuwait	12.7	108	—	—
USSR	8.0	13	—	6.8
Mexico	7.6	56	—	—
Iran	6.7	71	—	—
Iraq	6.3	75	—	—
Abu Dhabi	4.1	81	—	—
USA	4.1	9	—	28.0
Venezuela	3.6	39	20.0	—
Libya	2.8	55	—	—
China	2.4	19	—	—
Nigeria	2.2	30	—	—
UK	0.7	6	—	—
Indonesia	1.1	17	—	—
Algeria	1.1	27	—	—
Canada	1.0	12	19.3	—
Norway	1.4	31	—	—
Australia	0.2	7	0.05	0.8
Jordan	—	—	0.7	—
Morocco	—	—	—	7.4
Thailand	—	—	—	2.0

Largest 5 countries   57.7 (61%)
Largest 10 countries   78.6 (83%)

* Based on 1986 production
Sources: British Petroleum 1987; WEC 1983

of drilling rigs active outside the Centrally Planned Economies declined and the concentration in North America remained at approximately 65 per cent. Elsewhere during this period drilling activity increased only in Latin America (particularly Mexico).

Estimates of the amount of oil recoverable from shale and sands (Tables 6.1 and 6.3) have a different information base from those for conventional oil. In the first place they do not include estimates of expected discoveries but are limited to that proportion of known deposits occurring at depths accessible by open-pit mining (*in situ* recovery technologies are still in the early stages of development). Furthermore, for oil sands, they rely on estimated recovery rates based on first generation commercial scale operations in Canada and, for oil shale, upon non-commercial prototype operations in the USA. If potential geological resources were included and allowance made for improved recovery rates, estimates of the potential recoverable oil from each of oil shale and sands would considerably exceed that for conventional oil.

Spatially, over half the conventional oil reserves are thought to be located in the Middle East with one country, Saudi Arabia, containing almost 25 per cent of the world total (Tables 6.2 and 6.3). The potential geological resources are less concentrated with a little over half estimated to be shared between the Centrally Planned Economies (particularly the USSR) and the Middle East. In the USA, the UK and Australia the reserve/production ratio in 1986 was less than ten years and in the USSR, China, Canada, and Indonesia less than 20 years. At the other end of the scale the reserves in many of the Middle East countries and Mexico are shown to have lifespans of over 50 years.

Uranium   Estimating future uranium supplies is particularly difficult because of the relatively short history of exploration and the absence of a generally accepted methodology which recognizes the geological conditions in which exploitable deposits are likely to occur with some certainty (Edmonds and Reilly 1985). The result is estimates which are more like those for oil sand or shale in that they refer only to amounts of uranium in known deposits and extensions of those deposits. Consequently, future uranium supply data give little indication of the ultimately recoverable global resources which are thought by some to be as high as 80–250 million tonnes of uranium (Perry 1978). Estimates of the lifespan of uranium resources (Table 6.1) are also quite different from those of other fuels, because the amount required annually is particularly sensitive to developments in nuclear reactor technology (Table 6.4). Almost 70 per cent of the current reserves are located in five countries (USA, Australia, South Africa, Canada and Brazil) and over 60 per cent of the estimated additional resources (Category I) are in only three countries, Australia, Canada and South Africa.

Natural gas   Because of the association of natural gas with oil, estimates of the reserve and resource base of conventional gas are closely related to those for oil. The estimates for associated gas are made by applying assumed gas/oil ratios to the projections for oil. Non-associated gas estimates, however, are

163

Table 6.4 Uranium requirements of nuclear reactor types

Reactor type	*Tonnes U_3O_8 per 30 years at 70% capacity*	
	No recycle	*Recycle*
Light water reactor	6010	4750
High temperature gas reactor	4310	1770
Canadian deuterium	4380	1770
Liquid metal fast breeder reactor	Negligible	—

Source: Edmonds and Reilly 1985

made independently of oil but the information base is even more speculative than that for oil because of the relatively late emergence of gas-targeted exploration.

Global projections indicate a lifespan of almost a century at current levels of production (Table 6.1) although the meaning of this figure is clouded by the interpretation of 'production'. Production of associated gas occurs when oil is produced at a rate determined by the gas/oil ratio. Such gas may be marketed, flared or recycled back into the ground. In the latter use there is little net loss to the resource base. Spatially, current reserves are strongly concentrated in the Centrally Planned Economies and Middle East with over 50 per cent in just two countries, the USSR and Iran.

Coal   Projections of the global stock of coal establish it as the most plentiful conventional fuel. The dominance of coal is understated in Table 6.1, because the Potential Geological Resources category includes estimates for known fields only: yet-to-be-discovered coal-fields are not included. Consequently, as each new assessment appears global estimates increase (Edmonds and Reilly 1985).

On the basis of available information, stocks of coal are remarkably concentrated in three countries (USA, USSR, China) with substantial reserves in Poland, the UK and West Germany (Tables 5.18 and 6.6). As coal exploration (as distinct from delineation) activities increase, opinions differ on the likelihood of this basic distribution changing. Some think that new discoveries will result in a broader spatial distribution (Wilson 1980) whereas others are less optimistic (Edmonds and Reilly 1985).

The current state of knowledge of global non-renewable resource stocks may be very generally summarized as follows:

1. All estimates of future stocks are characterized by considerable uncertainty.
2. The lifespan of conventional oil is the least of the fuels.
3. The stocks of oil contained in unconventional sources are very large.
4. The stocks of conventional natural gas and uranium are moderate but substantial.

Table 6.5 Natural gas – projected reserves, reserves/production ratio and proportion of non-associated gas: selected countries

	Current reserves		Reserve/production ratio	Proportion Non-associated
	10^{12} cu metres	% world	years*	%
USSR	43.9	43	64	60
Iran	12.7	12	>100**	58
USA	5.2	5	12	72
Qatar	4.3	4	>100	90
Saudi Arabia	3.5	3	>100**	10
Algeria	3.0	3	86**	83
Norway	2.9	3	>100	N/A
Canada	2.8	3	40	89
Mexico	2.2	2	84	20
Netherlands	2.0	2	31	100
Venezuela	1.7	2	96	N/A
Indonesia	1.4	1	37	N/A
Malaysia	1.4	1	>100	N/A
Nigeria	1.3	1	>100**	N/A
Kuwait	1.0	1	>100	N/A
Largest				
5 countries	69.6	68	—	—
10 countries	82.5	81	—	—

* Marketed production
** Gross production
Source: IEA 1982; British Petroleum 1987

Table 6.6 Coal – estimated reserves and potential geological resources: World and regions

	Current reserves (10^9 tonnes coal equivalent)				Potential geological resources (10^9 tonnes coal equivalent)			
	Hard	Soft	Total	%	Hard	Soft	Total	%
Centrally Planned Economies	297	196	493	48	5,554	906	6,460	64
North America	136	135	271	28	1,286	1,399	2,685	26
Western Europe	35	61	96	9	401	18	419	4
Pacific	43	43	86	8	222	49	271	3
Other	68	4	72	7	265	27	292	3
World Total	579	439	1,018	100	7,728	2,399	10,127	100

Source: IEA 1982; British Petroleum 1987

5. The broad distribution of reserves of conventional fossil fuels is uneven with the majority of oil reserves in the Middle East, natural gas in the USSR and Iran, and coal in North America, USSR and China.

Future demand

Variables

The major variables affecting energy demand have been identified and briefly discussed in Chapter 2. In this section a selection of those variables is considered in terms of the uncertainties associated with projecting their future magnitude and influence at the world scale.

Population Demographic considerations play a basic role in assessing the medium and long-term outlook for energy demand in two ways. First, every additional person generates a small increase in the demand for goods and services and, thus, the energy required to produce them. Second, the magnitude of a population, in conjunction with its age structure, determines the size of the working force and, thus, the potential output of goods and services.

Population projections require the estimation of birth and death rates and, for individual areas, net migration rates. Projections of the global population made in the 1930s underestimated the actual figures because the death rate declined faster than expected. Post World War II projections, on the other hand, led to overestimates because the birth rate fell faster than was expected. Estimates made in the early 1980s (Keyfitz *et al.* 1983) project a doubling of world population to approximately 8×10^9 by 2050. A notable decline in the rate of increase in the next 25 years is expected to lead to an ultimate global population of $8.5-10 \times 10^9$. By 2050 the world's population of working force age is expected to increase 2.5 times or almost 80 per cent of the projected population. The developing countries' share of total and of labour force age population increases to over 80 per cent in these projections. Furthermore, the rate of urbanization in these countries is projected to remain high (Häfele 1981).

Economic activity The demand for energy largely derives from the level and structure of economic activity. The most commonly used measure of the level of economic activity at the national scale is Gross National Product (GNP) or the market value of the total output of goods and services during a specified period. The GNP results from the product of labour force and productivity. Estimates of future GNP values are made either on the basis of projections of these two parameters or by extrapolation of historical trends adjusted to take account of consistency requirements and judgements about the influence of changing circumstances. Among the latter are maintenance of environmental quality, political instability, capital availability and, emphasizing the interdependency between demand and supply, the price and availability of energy.

The relation between energy demand and GNP is commonly expressed by two measures: the amount of energy used to produce one unit of GNP (the energy/GNP ratio or energy intensity) and the percentage change in energy used in relation to the percentage change in GNP (the energy/GNP elasticity). Both of these aggregate measures represent the outcome of the interplay of a number of variables including the structure of the economy, prevailing technologies and stocks of equipment, and, once again, the price and availability of energy. Energy intensity ratios and elasticities vary over time and between countries (Cain and Nevin 1982) making specific values unsuitable for use in estimating global futures. However, estimates of future elasticities display a narrowing range of 0.4–0.8 for developed economies and 0.9–1.3 for developing countries (Edmonds and Reilly 1985). These estimates point to a smaller growth of energy demand per unit of GNP increase in the developed economies than in developing countries.

Price In the consideration of the effect of price upon energy futures it is useful to distinguish between the conditions which cause prices to change and the effects of those changes upon future energy demand and supply. In practice this is much easier said than done because of the considerable degree of interdependency between the two.

Price changes may be initiated from the supply or the demand side. From the former, price increases may result from production moving to higher recovery-cost reserves (e.g. to frontier or deep off-shore oil reserves), from governments increasing the administered element of energy prices (e.g. OPEC in 1972 and 1979), and from major interruptions of supply such as that caused by the Iran-Iraq war. Supply side influences may also serve to decrease the price if production exceeds market requirements (e.g. oil in 1986). From the demand side, upward pressure on prices occurs during periods of high economic activity and associated energy requirements or, in the shorter term, the occurrence of extreme temperature conditions in the heavily populated areas of the developed world. Low levels of economic activity and attendant diminished energy requirements serve to depress prices.

However induced, upward price changes tend to diminish demand and vice versa. The sensitivity of demand to price changes varies between energy sources, forms and uses and changes occur only after a time-lag the length of which also varies. The study of energy price elasticities receives considerable attention from economists but the results, while confirming the direction of change expected from first principles, unfortunately appear to have limited generality and applicability to prediction (UK, Department of Energy 1977a). Nevertheless the prediction of energy prices is probably a greater problem than dealing with the relationship between price and demand (Stern 1984).

Efficiency of use A fourth important variable is the efficiency with which energy is used. Three factors play a part in determining users' concern for efficiency: price, government programmes, and lifestyles and attitudes. As real energy prices rise, users become increasingly willing to make the capital investment and operational changes necessary to reduce their input, particularly if energy costs are a large proportion of their total costs. Government programs may range from mandatory regulations to incentives and information programs. The latter may be successful in changing user habits and fostering a 'conservation ethic'.

Methodology

In the most general terms the projection of energy demand involves:

(1) description of the magnitude and structure of current demand;
(2) identification of the variables which affect the current scene;
(3) calibration of the influence of each variable upon demand;
(4) estimation of the future magnitude of selected variables;
(5) projection of the effect that each of these variables will have.

These requirements are now commonly met by means of formal modelling using econometric, system dynamic and engineering process approaches (Stern 1984).

Econometric models, relying on regression equations to express the relationship between variables and available data, tend to produce projections based on the expectation that those relationships evident in the past will extend into the future. Systems dynamics models use partial differential equations to express the assumed relation of one variable to others in the recent past. However, even though systems models are capable of producing a variety of projected outcomes there is a danger that they '. . . can easily build more and more detailed pictures from less and less well documented knowledge' (Stern 1984, p. 10). Engineering process models lead to projections that embody economic and demographic assumptions but focus on factors such as the changing structure of energy using equipment and developments in energy efficiency (Schipper et al. 1985). Despite the potential strength of such models arising from their focus on technological change they too contain many assumptions about the rate and directions of such change and face difficulties in scaling-up efficiency data from test facilities to actual conditions in the everyday world.

Projections

The four demand projections summarized in Table 6.7 illustrate the range that exists between expert views of the future. The differences result from the conditions prevailing during the period during which the projection was undertaken and the influence of different assumptions and methodologies. The remainder of this section is devoted to a brief review of the most recent projection included in Table 6.7 (Edmonds and Reilly 1985).

The authors emphasize that the model used is a formal attempt to present a conditional, structured and consistent scenario which recognizes that there are

Table 6.7 Projections of total world energy requirements

	10^9 tonnes oil equivalent			
	WEC (1978)	Häfele (1981)	WEC (1983)	Edmonds (1985)
1972	5.5(A)	—	—	—
1975	—	5.8(A)	—	6.0(A)
1978	—	—	6.8(A)	—
1980	7.0	—	—	—
1990	9.5	—	—	—
2000	12.8	high 11.9 low 9.6	10.9	11.0
2020	22.7	—	16.9	—
2025	—	—	—	21.0
2030	—	high 25.1 low 15.8	—	—
2050	—	—	—	37.4

A = Actual
Sources: World Energy Conference 1978; Häfele 1981; World Energy
Conference 1983b; Edmonds and Reilly 1985

'. . . uncertainties at every turn' (Edmonds and Reilly 1985, p. 243). The model is intended as a tool to help in the assessment of the interaction of economic conditions, energy use and CO_2 emissions. The temporal scope of the projection is long term (to 2050) reflecting the nature of the CO_2 problem, energy is disaggregated into six component sources and the world is divided into nine regions. Five major variables are considered (population, economic activity, technological change, energy prices and energy taxes and tariffs), equations of several different structural forms are developed and specified for each module, and the major assumptions are made explicit all in the context of a supply–demand balance.

Edmonds and Reilly project that the rate of increase of world primary energy demand will decline to less than 2.5 per cent per annum (compared to 5 per cent p.a. 1950–75) in response to lower population and GNP growth and higher prices. Energy/GNP ratios are expected to remain relatively stable in response to counteracting tendencies of decline in the Developed Market and Centrally Planned Economies and to increase in the Developing Economies. Energy prices are assumed to increase rapidly up to the turn of the century, more slowly in the period 2000–25 and approach stability by 2050.

The mix of energy sources to meet the requirements is projected to change markedly: oil and gas down from 62 per cent in 1975 to 24 per cent in 2050, coal up from 29 per cent to 45 per cent and nuclear from 5 to 22 per cent (Table 6.8). In association with these changes, coal will replace oil in volume of trade. The distribution of consumption is also expected to change significantly. The share of the Developed Economies and USSR/Eastern Europe is projected to

169

Table 6.8 Projection of world primary energy requirements by source

	10^9 tonnes oil equivalent				Percentage			
	1975	2000	2025	2050	1975	2000	2025	2050
Oil	2.7	3.5	4.2	6.8	45	32	20	18
Gas	1.0	1.9	2.6	2.0	17	17	12	6
Coal	1.7	3.4	8.0	17.0	29	31	34	45
Hydro	0.6	1.3	2.7	2.7	9	12	13	7
Nuclear	—	0.9	3.6	8.3	—	8	17	22
Solar electric	—	—	—	0.6	—	—	—	2

Source: Edmonds and Reilly 1985

Table 6.9 Projection of primary energy requirements: World and major economic groups, 2000–50

	10^{18} J				Percentage of world			
	1975	2000	2025	2050	1975	2000	2025	2050
World	263	485	921	1646	100	100	100	100
Developed Market Economies	155	251	396	591	59	52	43	36
Centrally Planned Economies	79	135	258	484	30	28	28	30
Developing Economies	29	99	267	339	11	20	29	31

Source: Edmonds and Reilly 1985

decline from more than 80 per cent of global consumption in 1975 to 66 per cent in 2050 and that of the Developing Economies to increase from 11 per cent to 31 per cent with particularly large increases in China (which becomes the largest consumer) and Latin America (Table 6.9).

Summary

This chapter has dealt with the questions of how much and what form of commercial energy will be required in the future and from what sources and at what price it can be supplied. Answers to these questions are of such great importance to society, to corporations and to governments that they must be sought even though it is recognized that, over the long term, they have a high degree of uncertainty. Indeed, projections are not statements about what *will* happen in the future but rather about what *could* happen if specified conditions are met. They are carried out by modelling procedures which make use of only *selected* parameters and variables, have a strong dependency on being able to quantify the *relations* between the variables in the case of energy systems, and have to deal with *interdependencies*. Basically, future supply models seek to provide answers to three questions: how much exists, how much *can* be produced and how much *will* be produced?

In interpreting projections it is important to have a clear understanding of the intended *scope* (temporal, spatial and phenomenal) and the meaning of the *terms* used. In respect to future supply of mineral fuels, it is helpful to distinguish between *undiscovered resources* and *reserves* which may have conditional or current status. *Conditional* reserves may become current reserves if quality and size, technological, economic, and socio-political conditions (including concerns about environment and health) permit. *Current* reserves may revert to *conditional* if any of these circumstances become unfavourable. Technology is often singled out as an especially important and dynamic condition which influences what *can* be discovered, produced and used. Technological development involves four sequential stages (research, development, demonstration and deployment) which take considerable time to work through. Consequently, technological developments, fundamental as they are to energy futures, cannot have large-scale effects in the short or medium term.

Projections of future demand require consideration of a different set of variables and yet there is an unavoidable connection between future demand and supply. *Population* size and age structure is the most basic underlying driving force determining long-term energy requirements. Not only does each additional person require an extra increment of energy, however small, but as population increases so does the labour force and its influence upon *economic activity*. Both the level and structure of economic activity are major determinants of energy requirements influencing both the total amount and the mix of sources and forms of energy required. *Price* is the third important variable which will affect future energy demand and one which has a major influence on the fourth, *efficiency of use*. If overall energy prices rise, total energy demand will tend to fall (and vice versa) and if price differentials develop between energy sources the preferred mix of energy will change. Future energy demand projections are also arrived at by modelling which involves *description* of the current state of the demand system, *identification* of the variables that determine the state and *calibration* of their influence, *estimation* of the future magnitude of each variable and the *projection* of their influence on demand.

Projections of future energy requirements vary over a wide range depending upon when they are prepared, the variables considered and the values and influences assigned to them. All conventional wisdom accepts increased total energy demand in the medium and long-term future: there is, however, not a consensus on how great the increase will be or how rapidly it will occur beyond a general agreement that it will take place less rapidly than previously because of the likelihood of increased efficiency of use. Similarly it is widely held that the mix of energy sources which will be used to meet the demand will change and the relative role of conventional oil and gas will decline and the world will come to rely increasingly upon coal and uranium. Notwithstanding the expectations of the advocates of renewable resources and 'soft-energy paths', realistic projections agree that these sources will not make a significant contribution to energy requirements on a *global* scale (as distinct from a local or regional scale) for a long time to come. In terms of the three major economic groups, the relative spatial distribution of energy consumption is expected to become almost equal by the middle of next century provided the Developing Economies can acquire the necessary energy supplies.

The availability of future supplies to meet these requirements on a gross global basis indicates that, in the lifetime of most who read this book, the supply issue to be faced is that of conventional oil. The spatial concentration of reserves in the Persian Gulf area emphasizes the crucial importance of this area as Western Europe, the USA, Japan and, almost certainly before the end of the century, the USSR increasingly have to turn to the same area for supplies.

CHAPTER 7

Energy policy

Policy may be defined as the complex of objectives, strategies and programmes by which decision makers manage the system for which they are responsible. Thus there can be policy at the personal, corporate and government levels (Doern and Aucoin 1979). This chapter deals only with government or public policy, case studies of which are reviewed at three scales: national, international and global. As an introduction, however, it is useful to review the general structure of policy in order to provide a framework for the case studies.

The structure of policy

The large literature on public energy policy presents an array of disciplinary perspectives and substantive emphases. The latter may be summarized into a structure of policy consisting of the context, formulation, elements and results of policy (Fig. 7.1).

The context

The public management of energy systems at any scale develops in the context of information about the state of the system, the internal and external variables which influence the state, the values of the society concerned and the political disposition of the government in power. Acquiring an adequate information base about the current state of the energy system is itself a complex task facing such problems as the determination of the resource base and developing a comprehensive and ongoing knowledge of sectoral and end-use consumption patterns. As a result of the intensive socio-economic analysis of energy systems which commenced in the 1960s, the identification of the important variables is not difficult for most governments. However, quantification of their influence

173

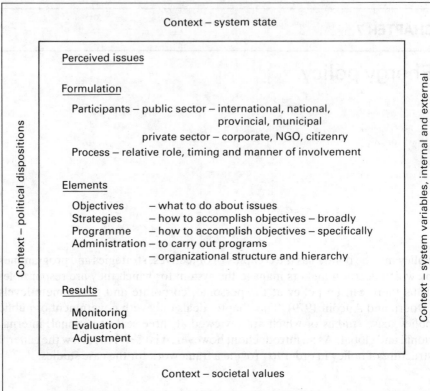

Context – system state

Perceived issues

Formulation

Participants – public sector – international, national, provincial, municipal

private sector – corporate, NGO, citizenry

Process – relative role, timing and manner of involvement

Elements

Objectives   – what to do about issues
Strategies   – how to accomplish objectives – broadly
Programme   – how to accomplish objectives – specifically
Administration – to carry out programs
                  – organizational structure and hierarchy

Results

Monitoring
Evaluation
Adjustment

Context – political dispositions

Context – system variables, internal and external

Context – societal values

Figure 7.1 A structure of policy

upon the system at useful levels of accuracy is still an elusive goal and prediction of the future values of the variables (particularly those external to the political unit) and their influence is fraught with uncertainty. Despite the information limitations, governments which seek to manage the energy systems in their jurisdiction must develop policies and programmes which will achieve the desired objectives and yet are sufficiently flexible to take account of the unavoidable inaccuracies and uncertainties of the information base available to them.

Prevailing socio-economic values, the political ideology of the government in power, and the nature and extent of regional differences constitute other components of the context within which public policy is set. Values and ideologies determine to a considerable extent the way in which the structure and performance of the energy system is interpreted, the issues that are perceived and the strategies and programmes that are implemented. As a consequence of these contextual variables, public policy may be subject to relatively frequent and drastic change.

Perceived issues

As already noted the energy context of a country as documented by the available data is subject to differing interpretations depending upon the value system and circumstances of the government concerned. Thus the political philosophy of the government of the day and the political, economic, and social conditions in the area governed determine what aspects of the national energy system are perceived as issues. Generally, governments of producing areas perceive the basic issues to be centred on (1) establishing jurisdiction over the resources, (2) the production and marketing of the resources, (3) the amount and distribution of the resource rent and (4) planning for the consequences of resource depletion. On the other hand, governments of net consuming areas perceive the primary issues in terms of (1) security of supply, (2) cost of supply and (3) amount and efficiency of consumption (Hughes *et al.* 1985, pp. 56 ff).

Policy formulation

The formulation of public policy may be considered on the one hand in terms of the participants involved and on the other, the manner or process of their involvement. In addition to the executive, legislative and administrative arms of government (at both the national and sub-national level), potential participants include private sector corporations and industrial associations, research organizations and the citizenry at large. The latter may participate as the electorate and as action-oriented special interest groups.

The relative role assigned to these participants in public policy formulation and the timing and manner of their involvement vary widely. There are clear differences between the process in countries with centralized governments and those with Western style democracies and even between the latter there is considerable variation. In some, citizen action groups are formally involved, in others their role is very limited; royal commissions are commonly used in Commonwealth countries, permanent legislative committees in, for example, the USA; in some cases private energy sector voices are dominant, in others they are unheard.

The elements

For the purpose of analysis the substantive content (or elements) of public policy may be grouped under four headings. The first, objectives, defines the directions in which policy is pointed and the last three (strategies, programmes and administration) deal with aspects of implementation (Smit, Johnston 1983).

Objectives Public policy in producing areas seeks to establish sovereignty over the resource and jurisdiction over its production, to determine its size and quality, to expand the market and to establish a fiscal structure which maximizes resource rent to government. In net importing areas, supply side objectives frequently include the determination and mobilization of indigenous resources, the diversification of the sources from which imports are drawn, protection against supply interruptions, and acquisition of the least cost, environmentally benign sources. On the demand side, desired shifts from one fuel to another and efficiency objectives are identified in total and for each consuming sector.

Strategies Realization of the specified objectives requires first the development of broad strategies and then of programmes and organizations to implement them. Some major strategic considerations include the relative importance to be attached to the energy sector in relation to other economic sectors, the relative role to be assigned to 'market' and 'interventionist' forces, the degree of flexibility to accept in the face of the uncertainty arising from uncontrollable variables, and the time period appropriate for policy development. Choices among these and other considerations are made by governments on the basis of political ideology and the prevailing wisdom about the behaviour of the national economic system. The political system of most democratic countries which require governments to go to the polls every five years or less can result in major changes of strategy (indeed of the entire structure of policy) over relatively short periods. In these circumstances, and considering the uncertainties of energy system behaviour, governments tend to adopt reactionary (short term) rather than anticipatory (long term) strategies.

Programmes The programmes used by governments to achieve their policy objectives may be broadly divided into three classes: regulatory, fiscal and participatory. Regulatory programmes are widely used to control exploration, production, emission and disposal of wastes, safety, trade and many other elements of the energy system. Fiscal programmes are used to shape aspects of the supply side of the energy economy, to control the magnitude and distribution of resource rent as well as to influence consumption patterns. Direct participation involves the establishment of government controlled firms in the energy industry or the assumption of a major equity position in existing private sector corporations.

Administration Public energy policy requires an administrative structure both to assist in its formulation and to carry out the implementation phase. Many governments form line ministries or departments of energy often with associated regulatory and advisory bodies. Others do not favour a single comprehensive energy department but, rather, have separate organizations responsible for individual energy sectors (e.g. solid fuels, hydrocarbons, hydro and nuclear) and, within them, divisions responsible for different stages such as resource evaluation, primary production, etc. Even with the single department of energy approach, many other government departments are involved with one or more aspects of the energy system. Thus, whether the centralized or dispersed model of organization is favoured there is a network of inter-departmental linkages which must operate in concert if policy is to be implemented effectively.

Policy results

After the energy policy is articulated and in operation, questions of effectiveness and appropriateness arise. In order to answer these questions monitoring and evaluation must be carried out both internally and externally. Internally, monitoring and evaluation may be carried out continuously by line agencies established explicitly for the purpose, by multi-department working groups, by quasi-independent boards or by internal committees formed periodically. More elaborate and less frequent reviews and evaluations are carried out by commissions and task forces made up of government appointed, but external, members.

External monitoring and evaluation is also conducted by scholars and commentators acting individually or as members of independent policy analysis organizations. Most developed countries have one or more universities with energy research centres in which all or part of their work is devoted to policy analysis.

National policy

The domain of public energy policy extends only as far as the jurisdiction of the level of government concerned. That jurisdiction with respect to the various parts of the energy system is determined by the constitutional structure of government. In some countries the division of powers between the state (province) and national levels of government poses serious problems for the development and implementation of policy. The following reviews of the energy policy of the USSR, USA and Japan are set at the national level and focus on only the major objectives and associated programmes (Table 7.1) with a concluding commentary on results.

USSR

In 1987, the USSR had the world's largest reserves of natural gas, very substantial reserves of other fossil and nuclear fuels and led the world in production and export of commercial energy. The spatial extent of the country, while affording considerable resource opportunities, also presents formidable transportation problems. Much of the newly developed and potential energy resource base occurs in harsh (and sensitive) Arctic and sub-Arctic environments which make all components of the energy production chain difficult and costly.

Objectives and programmes Successive Soviet leaders have explicitly recognized the importance of energy to the evolution of the USSR economy and most Five Year Plans have included large-scale energy initiatives. The persistent underlying objectives of these programmes have been to develop energy production and the required infrastructure to a level sufficient to supply the

177

Table 7.1 Selected elements of the energy policies of the USSR, USA, and Japan

Country	Objectives	Strategies and programmes
USSR	1. Increase energy production	– in 1986–90 increase production of natural gas 30%, electricity 20%, coal 10%, and oil 3%
	2. Increase efficiency of use	– decrease overall energy consumption
	3. Increase (or at least, maintain) exports to hard currency areas	– substitute other energy sources for oil to leave maximum supply for export – increase exports of gas, coal and maintain oil
USA	1. Foster ENERGY STABILITY to prevent problems of availability and price from destabilizing economy and way of life	– deregulate energy industry and maximize role of free market forces – focus public R & D effort in selected areas to 'proof-of-concept' stage – support programmes to stabilize international scene
	2. Foster ENERGY SECURITY so that adequate supplies from domestic and foreign sources are physically available at reasonable cost	– maintain emergency preparedness via Strategic Petroleum Reserve – encourage development of domestic sources – open Federal Lands and Outer Continental Shelf to exploration and production
	3. Foster ENERGY STRENGTH by developing a balanced and diversified mix of supply sources and conservation	– focus on the development of the 'energy triad' of conservation, coal and nuclear energy – conservation: give free play to market forces and prepare user guidelines, not regulations – coal: increase access to resources on Federal Lands, expedite transportation developments and support R & D to diminish environmental impact of combustion – nuclear – reform licensing procedures, pursue waste disposal programmes, remove barriers to trade and restore US primacy in uranium enrichment – other – accelerate decontrol of gas industry
Japan	1. Secure stable flow of energy imports	– diversification of supply areas – stockpile, promote international cooperation
	2. Decrease dependence on oil	– increase use of coal, uranium, LNG – increase R, D & D on both supply and use
	3. Increase efficiency of energy use	– reduce energy-intensive industries – replace equipment – improve operating practices

178

needs of rapid growth in the Soviet economy, to support an export trade to hard-currency, technologically advanced economies and to supply the CMEA countries of Eastern Europe.

The 12th Five Year Plan (1986-90) continues to reflect these broad objectives. Under the leadership of M. Gorbachev, the Soviet government has developed specific goals and priorities which take into account their perception of the state of the national economy and, with particular reference to energy, the changing global scene. The overall economic strategy expressed in the 1986-90 plan emphasizes the importance of increasing the efficiency of the economy by upgrading and more intensive use of existing facilities. Growth is to be achieved mainly by increasing productivity rather than expanding the production base into new areas and facilities. A key component of this strategy is the reduction of overall energy intensity by 7 to 9 per cent during the life of the plan (Kaser 1986).

Despite energy savings, increased energy supplies are required. The 12th plan proposes that the production of natural gas, electricity and coal should be increased over that in 1985 by approximately 30, 20 and 10 per cent respectively Crude oil output is to be raised by 2 to 3 per cent and sustained at that level until the end of the planning period (Shabad 1986). To meet these targets the plan provides for intensification of production within developed resource areas by such means as secondary and tertiary oil recovery, increased utilization of associated gas and the enlargement of existing open-pit coal mines. In addition to these intensification measures some extension of both exploration and development activity into more remote areas of Siberia is planned. Increased production of electricity is to be accomplished in the western USSR mainly by the expansion of existing nuclear plants and developing more pumped-storage facilities. East of the Urals, the emphasis is to be on expansion and new construction of gas and coal fired plants.

Programmes to maintain or increase energy exports to hard-currency areas take several forms beyond simply those designed to increase production. In order to make the largest possible proportion of total oil production available for export, the 12th plan provides for a continued and accelerated reduction in the use of oil in the national economy. In particular, gas is to be substituted for oil as a general boiler fuel and gas, nuclear fuels and coal for oil-fired central electric stations. With spare capacity in trunk pipelines and in the absence of near-term resource limits or major production difficulties, increased natural gas exports will be pursued by aggressive marketing in Western Europe, extension of export pipelines to Greece and Turkey and continued monitoring of market opportunities in Japan (Gorst 1987). Additional sales of coal from the Soviet Far East to Japan will also be pursued (Bradshaw 1987).

The general Soviet strategy of fostering and guiding the development of the satellite economies of Eastern Europe can, in part, be accomplished by means of energy policy. By providing these countries with energy (particularly hydrocarbons and, in some cases, electricity), the USSR assists with their economic development and, by tying such supplies to performance require-

179

ments, exerts some control over the direction of development (see CMEA section below).

In order to achieve the targets identified in the 1986-90 plan very large allocations of capital will be required along with organizational and attitudinal changes in management. In the 11th Five Year plan the energy sector was allocated more than 80 per cent of the incremental investment in all industry (Gustafson 1983). In the 12th plan it is proposed to increase the allocation (particularly to the oil sector) so that energy-related expenditure will constitute over one-third of total industrial investment (Kaser, 1986). There is little information available about the specific steps proposed to overcome the organizational and attitudinal constraints in the energy sector noted earlier in the decade (Dienes 1981). Generally, however, the overall context of economic reform and shifts in decision-making roles toward the production agencies suggest some significant re-assignment of responsibilities (Kaser 1986).

Results In mid-1987 the 12th plan was still in its early stages and it was not possible to judge the results with any certainty. By 1986, oil production and exports had almost returned to 1983 levels but informed opinion remains divided on the constraints facing the Soviet oil industry and its ability to increase production (Gustafson 1985; Gorst 1987; Wilson 1986). Although the haste with which Western Siberian gas fields have been developed has led to some wasteful practices which in the long term may have serious implications, market rather than supply constraints appear to be of more concern for gas in the short term (Kroncher 1985). This not only refers to export markets but, to some extent, to the development of domestic markets which is constrained by the ability to carry out retrofitting at existing plants (Sagers and Tretyakova 1986).

The major uncertainty beyond the performance of the oil industry is the effect of the failure of the Chernobyl nuclear plant in 1986. Whether this large-scale accident will slow down the drive to increase the role of nuclear generation of electricity in the USSR is still unclear. Whatever the final outcome, Soviet policy makers will '. . . have to re-evaluate plans and reconsider the technological structure of future power capacity' (Thornton 1986, p. 15) thus making the planned 20 per cent increase of electricity production all the more difficult to obtain.

USA

In the first half of the twentieth century, the United States was the storehouse of the world's established fossil fuel reserves, the major world supplier of oil, and the largest and most profligate consumer of energy. By the latter part of the century, established reserves of coal are still large, but those of oil and gas are declining. As a result, the United States has become an increasingly large importer of oil such that in 1977 almost half of its total crude oil consumption was imported. After a decline in the early 1980s, this figure is expected to increase again by the end of the decade (US, DOE 1985, p. 41).

Objectives and programmes   By the late 1970s energy matters had emerged as the major domestic issue in the United States. The Nixon and Ford administrations had started to develop an explicit energy policy with the establishment of emergency authorities and increased public funding to encourage alternative energy sources. The Carter administration, in 1977, developed a large-scale, comprehensive programme that was highly interventionist and regulatory in style. In 1981, under the Reagan administration, the relative importance of energy was downgraded so that it was no longer seen as a discrete problem '. . . but a policy issue to be treated as part of an overall economic recovery program.' (Parker *et al.* 1981, p. 15). This administration introduced a policy plan which transferred '. . . the focus of decision making from the Federal government to the States, to the private sector and even to individuals.' (Parker *et al.* 1981, p. 1). Further, it was to rely upon '. . . market oriented policies that build upon America's vast production and conservation resources and its technological genius, free of arbitrary regulation . . .' (US, DOE 1985, p. 3).

Within the overall goal of providing an adequate supply of energy at reasonable cost the National Energy Policy Plan (NEPP) of 1985 identified energy stability, security and strength as the basic objectives. Energy stability refers to the prevention of availability and price problems from destabilizing the United States economy or way of life. This objective is to be achieved primarily by the deregulation of the energy industry and, within a reduced total allocation of funds to research and development, a focused programme of basic research into the areas most likely to lead to commercial development by the private sector.

Energy security is to be achieved by the provision of adequate supplies at reasonable cost from either domestic or foreign sources. Thus, unlike earlier administrations, the reduction of imports *per se* is no longer a national objective. Energy security programmes include the maintenance of energy preparedness by building the Strategic Petroleum Reserve (SPR) and encouragement of the establishment and production of domestic resources by increasing accessibility to the Federal Lands and Outer Continental Shelf (OCS).

Achievement of both energy stability and security is expected to contribute substantially to the third objective, energy strength, which is seen to require increased attention to conservation and the building of a balanced and diversified mix of supply resources. The 1985 NEPP refers to a triad of conservation, coal and nuclear energy as playing a leading role in achieving energy strength while recognizing the contributions from further deregulation of the gas industry and the opening of the OCS to exploration (US, DOE 1985, pp. 13 ff).

Continued improvement in the efficiency of energy use (conservation) is expected to arise from the free play of market forces and the preparation and dissemination of voluntary guidelines for users. The increased use of coal is not considered to have resource-based constraints (though if these arise, increased accessibility to the western Federal Lands will remove them) but, rather, is inhibited by environmental impact and, to a lesser extent, transportation problems. Public and private research programmes focused on combustion technology are expected to reduce the impact problem to acceptable levels and a new and decreased regulatory structure is intended to diminish the transportation constraints. The nuclear industry is to be encouraged by the streamlining of licensing procedures, development of a national nuclear waste disposal program, increased public funding for research focused on safety and an associated public information programme and the restoration of US primacy in uranium enrichment.

Results The general reliance upon market mechanisms and the repudiation of government intervention by the Reagan administration has been accompanied by at least short-term changes in the United States energy system which are in keeping with policy objectives. Overall energy consumption has declined, energy efficiency has increased, oil imports have declined and their sources have diversified, the role of oil in the energy mix has diminished, the Strategic Petroleum Reserve contained 500 million barrels in 1986 (Marshall 1986) and gas was still plentiful, at least up to 1987.

Just what proportion of these developments can be attributed to United States policy as opposed to more general economic circumstances and forces beyond the jurisdiction of the United States government is a moot point. Supporters of the current policy argue that the way in which changes have generally occurred in keeping with policy objectives justifies the government's faith in the operation of market forces. Whether this stance will be in the public interest over more than the short term, however, remains to be seen. Critics of the NEPP draw attention to a growing number of existing and potential problems (Editors 1985; Hirsch 1987; Katz 1984). Philosophically, opponents of the free market approach argue that it ignores the complexity of consumer behaviour and does not take sufficient account of social and institutional variables. Furthermore, economic rationality may operate at the expense of national security, and when prices are low (e.g. 1986 and 1987) the market sends signals which encourage neither efficiency of use nor expansion of exploration.

Other reactions are centred on environmental issues and the dangers associated with increasing oil imports. For some, the major environmental issue is the atmospheric impact of coal combustion and, in particular, the growing 'acid rain' problem. For others, it is the disposal of nuclear waste or the wildlife and recreational consequences of opening more Federal Lands and the OCS to exploration and, later, production of fossil fuels. These concerns clash with the NEPP proposals to shift to more coal and nuclear fired electricity plants and the maintenance of hydrocarbon reserves in formerly protected areas. Those who fear what they regard as the inevitable increase of oil imports argue for import quotas and duties and the provision of tax incentives for domestic exploration and production.

Japan

During the period of global economic growth following World War II, Japan emerged as one of the largest energy-consuming nations of the world with one of the highest oil dependencies of any country. With only very limited indigenous energy resources Japan rapidly became the largest importer of oil, coal and natural gas. This energy profile makes Japan very sensitive to changes in the international energy scene and an attractive market opportunity for energy exporters.

Objectives and programmes The over-riding objective of Japanese energy policy is to develop and maintain a stable flow of energy imports sufficient in amount and at a cost which will permit the realization of broader economic and social goals. Subsidiary, but still major, objectives focus upon inter-fuel substitution to reduce the dependence on oil and upon decreasing the general intensity of energy consumption in all sectors (Quo 1986).

Programmes aimed at diversification, international cooperation and stock-piling are both directly and indirectly supported in pursuit of the primary objective. Diversification is important in two senses. First, diversification of suppliers (particularly of oil) so as to minimize the effect of interruptions of supply from any one area and to give Japan as a buyer some leverage among suppliers. Second, diversification of the total energy supply mix away from the heavy dependence on oil. In both of these senses diversification is essentially a medium-term strategy and in the short term international cooperation and stock-piling are more effective. Particularly with respect to oil, Japan has developed its own stockpile and actively seeks to arrange access to the stocks of other producer and consumer nations.

The dominance of imported oil in the national energy system, an advantage during the period of low real prices, in today's uncertain oil supply scene leaves Japan highly vulnerable to both physical interruptions of supply and price escalations. Consequently, a number of initiatives have been taken to substitute other fuels for oil. In the medium term these initiatives are directed towards increasing the use of coal, uranium and natural gas (LNG) and in the longer term the introduction of unconventional technologies such as photo-voltaics and fuel cells (IEA 1986, 319–37).

In order to achieve the third major objective of further increasing the efficiency of energy use, Japan has recently embarked upon more ambitious programmes to conserve energy. These programmes are particularly directed at manufacturing (which is more dominant as an energy-consuming sector in Japan than in most developed economies) although programmes for the residential and commercial sectors are being accelerated (IEA 1986, 319–37).

Results On balance Japan appears to have managed its energy system successfully in that uncertainties, interruptions and rapid price rises of oil have been absorbed with less effect on the economy than might have been expected. There are, however, constraints to at least the timing of the achievement of

183

their major energy objectives. They remain vulnerable to any prolonged interruption of supply from the Persian Gulf and, despite the buyer's market of the mid-1980s, the major suppliers of coal, natural gas and uranium may baulk at too great downward pressure on prices. The off-oil programmes will require very large capital investment in new energy conversion equipment, transportation and storage infrastructure which even the Japanese economy may find difficult to mobilize.

Finally, because Japan has already made effective progress in decreasing the overall energy intensity of its economy (Hough 1987) further progress will take longer to realize. Aside from government assisted programmes to improve efficiency among small businesses and in the residential sector, it is recognized that further major improvement in energy efficiency will have to come from longer-term structural changes in the industrial sector (e.g. the migration of energy-intensive industries off-shore) and the introduction of new industrial processes.

International policy

International organizations are initially formed in order to promote the self-interest of their members. Some catalytic event or sequence of events leads the governments of individual countries to see advantages in cooperation and, in response to the initiative of one or more countries with a particularly strong self-interest, they coalesce to form an international organization.

Some of the earliest examples of international cooperation in energy emerged as the result of the advantages perceived by contiguous countries in building and operating joint electrical systems (Sewell 1964). With the politicization of energy affairs and the attendant growth of government intervention, international organizations have outgrown the purely operational function and now seek more complex objectives. Today there are three major organizations each of which constitutes the 'energy arm' of a more broadly conceived international group:

1. The Organization of Petroleum Exporting Countries (OPEC) which, in part, grew out of the Arab League.
2. The International Energy Agency (IEA) of the Organization for Economic Cooperation and Development (OECD).
3. The Long-term Target-oriented Program of Cooperation – Fuel, Energy and Raw Materials (LTPC–FERM) of the Council of Mutual Economic Assistance (CMEA).

The non oil-exporting Less Developed Countries (LDC) in some respects constitute a fourth group but they do not have formal, energy-focused organizations beyond those formed within such United Nations regional organizations as the Economic Commission for Latin America (ECLA) and the Association of South East Asian Nations (ASEAN). This section treats energy policy at the international level by reviewing the objectives and programmes of OPEC, IEA and the Eastern European members of the CMEA (Table 7.2).

OPEC

The Organization of Petroleum Exporting Countries (OPEC) was formed as the result of two separate initiatives which quickly converged. Immediately after World War II leaders of the Arab countries of western Asia formed the Arab League as a means of advancing the general interests of their countries in the postwar period. The potential importance of oil as the foundation of future economic development was recognized by the establishment of the Department of Oil within the League. Concurrently, Iran and Venezuela were in consultation about future oil policies, and they were quickly joined by the Persian Gulf Arab oil states led by Saudi Arabia. In 1960, representatives of Iran, Iraq, Kuwait, Saudi Arabia and Venezuela agreed to establish OPEC as a formal international organization which was duly registered with the United Nations in 1962. Subsequently the membership expanded to include 13 countries (Table 7.2).

Table 7.2 Selected elements of the energy policies of the OPEC, IEA and CMEA

Organization	Objectives	Programmes
Organization of Petroleum Exporting Countries (OPEC)	1. Maximize benefits from hydrocarbon resources	– national companies and concerted Organization action
	2. Maintain (increase) real income	– control price and production – protect value of investments in developed economies
Members: *Algeria* Ecuador Gabon, Indonesia Iran, *Iraq* *Kuwait, Libya* Nigeria, Qatar *Saudi Arabia* *United Arab Emirates* Venezuela (OAPEC members italic)	3. Broaden structure of and modernize economies	– participate in all stages of hydrocarbon industry – develop social and economic infrastructure
	4. Initiate global co-operation on economic affairs	– advocacy of and participation in international conferences and organizations
	5. Assist economic development of other developing countries	– fund for international development
International Energy Agency (IEA)	1. Coodinate non-communist, industrialized nations' responses to global energy developments	– standing group on long-term cooperation; annual review; meeting of energy ministers

continued

185

Table 7.2 (Continued)

Organization	Objectives	Programmes
Members: Australia, Austria Belgium, Canada Denmark, Germany Greece, Ireland Italy, Japan Luxembourg Netherlands New Zealand Norway Portugal, Spain Sweden Switzerland Turkey United Kingdom United States	2. Increase energy efficiency	– major comparative study to assist in selection of cost effective programmes
	3. Increase role of environmental consideration in policy development	– IEA Environment Committee and publications
	4. Decrease dependence on oil	– encourage substitution of other fuels and renewable resources for oil
	5. Decrease vulnerability to energy supply interruptions	– emergency oil allocation scheme – diversification of energy supply sources
	6. Increase indigenous energy production	– maintain free market conditions and appropriate price structure
	7. Maintain research and development investment and enlarge scope of international collaboration	– IEA Research and Development Committee – major review of cost-effective programmes
Council of Mutual Economic Assistance (CMEA)	1. Reduce negative energy balance with USSR	– maximize domestic production and trade between European countries
	2. Increase energy efficiency	– change structure of industry – replace and modernize equipment – improve operating practices
Members: USSR, Cuba Bulgaria Czechoslovakia GDR, Hungary Poland, Rumania	3. Reduce dependency on oil	– increase relative role of coal, nuclear, hydro and natural gas – R & D to support objectives 2 and 3
	4. Diversification of sources of oil imports	– bilateral and multilateral agreements outside CMEA

The central importance of the Arab countries in OPEC and the continuation of the Arab League initiatives is reflected in the formation, in 1968, of the Organization of Arab Petroleum Exporting Countries (OAPEC) consisting now of ten members. All but Bahrain, Egypt, and Syria are also members of OPEC, the ratified resolutions of which are binding on all OAPEC countries.

In 1962, few realized that OPEC was to become the organization that would revolutionize the structure of the oil industry and thus emerge as a major force in world affairs. Indeed after 1973, most observers expected that the life of the organization would be short because of the political, cultural and economic heterogeneity of its members. Despite the stress created by these differences, the lure of oil income and the crucial role it plays in socio-economic development has proved sufficiently strong to maintain a degree of cohesiveness which, although constantly tested, has surprised the world.

Elements of current policy (Table 7.2)   The initial objective of OPEC was to assume a significant role in determining the price structure of crude oil. This important, but limited, objective quickly expanded to include sovereign control over the resource and its production, direct national participation in the oil industry and increased presence in global affairs (Al-Chalabi 1980).

Within the over-riding rubric of maximizing benefits from their hydrocarbon resources, the most evident and persistent objective of OPEC is to maintain and, if possible, expand, the real income of member countries from the export of oil and, to a lesser extent, gas. To accomplish this objective, prices are set at intervals in relation to oil quality, spot prices in major world markets, the price of marginal supplies of oil, inflation rates and the prices of the goods and services imported by OPEC members (Ghosh 1983). Those countries that have been able to maintain a strong positive balance between income and expenditure, and thus have capital funds to deploy, have the additional objective of investing those funds where they will yield the highest returns. Indeed, for OPEC members in this position the placing and protection of off-shore investments may become a more important objective than short-term increases in oil prices or volume of sales.

All the countries in OPEC are in various stages of broadening and modernizing their economies. To a greater or lesser extent, they all seek to accomplish these changes by direct participation in both upstream and downstream development of the hydrocarbon industry and by establishing energy intensive manufacturing. This means entering the oil refining and gas processing industries (in either the home country or off-shore), as well as the petro-chemical industry and such energy-intensive industries as aluminium reduction (Fesharaki and Isaak 1983).

The third and fourth objectives of OPEC are outward-looking in that they seek to establish the organization as an orderly force for change in global affairs. In 1975, the OPEC summit meeting formally declared support for a dialogue on energy between the organization and the industrialized world, with the condition that such a dialogue would include North-South issues in general (Maull 1984).

Embodied in the general objective of initiating and participating in a broad global dialogue on the changing economic order, is a fourth, separately identifiable, objective of assisting the economic development of both OPEC and non-OPEC Less Developed Countries. To this end, a Special Fund (now the Fund for International Development) was established by means of which a consolidated approach could be taken to providing financial assistance to such countries. This fund provides capital in addition to the bilateral assistance provided by member countries.

In a broader way OPEC also seeks to provide leadership in developing a negotiating position for LDC in global affairs. This position includes concern for the stabilization of the prices of LDC commodity exports, food pro- grammes, transfer of technology, construction of energy processing facilities and protection of the value of OPEC investments in developed economies.

Between 1982 and mid-1987 the cohesiveness of OPEC was severely tested and, at times, failed. In the face of first a declining and, more recently, a soft market, increased production by non-OPEC countries and the development of spot and futures markets in oil, the world role of OPEC declined. The tensions that developed in the organization have been very divisive as each OPEC member has sought to maintain its income from oil sales (Aharari 1986). By mid-1987, a fragile production quota system was in place after numerous unsuccessful attempts during the period to control production or prices or both. The system served to return world crude prices to the US $18–20 barrel range after they had plunged to US $10–12 barrel earlier in the year (Ali 1986; Shwardran 1986).

IEA

The formation of the International Energy Agency (IEA) was initiated by the USA in response to the sharp price increases and selective embargoes imposed by OPEC in 1973. The Agency was formally established in 1974 within the Organization for Economic Cooperation and Development (OECD), and in 1987 the member countries include the major energy-consuming nations of the non-communist world (Table 7.2).

Elements of current policy In 1986 objectives of the IEA emphasized the over-riding aim of fostering cooperation among the major non-communist energy-consuming nations of the world in response to global developments. This pervasive objective is from time to time focused into specific goals in response to changing conditions. In the mid-1980s, the communique of the. Governing Board of Ministers emphasized items listed in Table 7.2 (IEA 1986).

The role that increasing efficiency plays not only in reducing total energy demand but in contributing to energy security and minimizing undesirable environmental consequences continues to be stressed. For member govern- ments to develop effective programmes, governments are encouraged to ensure that appropriate pricing and tariff systems are in place, standards and regulations are set and energetically applied and significant funds are allocated to information systems and the development of new technologies.

The environmental consequences of all links in the energy chain are assigned a high priority in the mid-decade Ministerial statement (IEA 1986). It is recognized that many factors play a part in the formulation of energy policy and that protection of the environment is now inescapably one of them. The IEA recognizes that not only must environmental impacts and associated mitigative measures be seen as unavoidable but they must be considered at an early stage in the shaping of policy and the approval of projects. The control of air pollution from fossil fuel combustion and the handling of nuclear wastes are specifically identified as issues requiring urgent attention.

Decreased dependence upon oil has been a major objective of the IEA since its establishment. As a result of the events of the 1970s the relative importance of oil has declined in all member countries in response to more or less vigorous 'off-oil' programmes. In many cases this has been achieved by the substitution of natural gas, coal and nuclear fuels for oil which has led the IEA, while still supporting the off-oil objective, to focus its attention upon the implications of these supply shifts. For instance, the dangers of becoming too reliant upon any one source of natural gas are noted and member countries are encouraged to diversify their suppliers, preferably among OECD countries.

Improvement of the security of oil supply to member countries was one of the founding objectives of the IEA. Although the dominance of west Asian suppliers has been considerably reduced, in the first half of the 1980s they remained major sources which if cut off by military or political events could cause considerable disruption. Consequently, the IEA not only encourages its members to take steps to maintain substantial reserves but itself has developed, and continues to test and improve, an organizational procedure for emergency sharing.

As a further protection against supply interruptions resulting from external events, the Agency urges member countries to pursue the development of indigenous energy resources. In the general context of slackened demand and lower prices of energy in the mid-1980s, the IEA is concerned to note that both the public and private sectors are delaying exploration and development investments in the expectation of a persistence of these conditions. Such a tendency, they observe, is a normal commercial response to changing market conditions but governments should be aware of the medium- and long-term implications of such delays and consider adjusting their policies to discourage any further slowdown of investment.

In 1985, and again in 1986, the IEA emphasized the need to strengthen research and development programmes in the respective countries, encourage the more rapid diffusion of results into the operational sphere and facilitate greater collaboration both between the public and private sectors and internationally between countries. As with the maintenance of investment in exploration and new projects, the slow growth of demand and the soft prices of the

1980s have discouraged the large-scale allocation of funds to research. The IEA is attempting to warn their members of the dangers of such a tendency.

Finally, it is interesting to note the Agency's position on the relative role of the public and private sectors. The general orientation is indicated in a 1985 statement which stresses:

> . . . that one requirement for energy policy is flexible, open and resilient markets within which the different fuels can compete vigorously and find their levels of demand and supply through operation of the price mechanism (IEA 1986, p. 92).

but:

> . . . governments must also carry out policies which accomplish their objectives and at the same time supplement an effective market system in a manner which is consistent with the national circumstances of each country . . . (IEA 1986, p. 93).

These two quotations make clear the difficulty of the task facing individual governments as they attempt to steer a course between too great or too little intervention. The public–private sector relationship is further complicated in an international setting such as the IEA which also faces the difficult task of trying to achieve concerted action by governments whose agendas are greatly influenced by their own circumstances and considerations of sovereignty and comparative advantage.

CMEA – Eastern Europe

Despite large-scale production of hard coal in Poland, lignite in Eastern Germany and moderate to small-scale production of other energy sources in the remaining countries, Eastern Europe is energy deficient (Chap. 3). Consequently, the underlying objective of energy policy of Eastern European countries is to secure a reliable supply of least-cost energy to sustain and enhance their economic progress. Externally, their choices are limited because of their general inability to earn sufficient hard currency to purchase energy in the world market and because of the spatial and political realities of their relationship with the USSR. Arising from this relationship, the countries of Eastern Europe and the USSR are grouped into the Council of Mutual Economic Assistance (CMEA). The Council of Ministers of this Soviet-dominated organization coordinates the overall economic policy of member countries by means of Long-term Target-oriented Programmes of Cooperation, one of which deals with Fuel, Energy and Raw Materials (LTPC-FERM). Subordinate agencies such as the Complex Programme for Energy and the Standing Committee on Electrification carry out programmes.

Elements of current policy The agreement on energy reported upon by mid-1987 arose out of a meeting of the Council in 1984 (Davey 1987). In this agreement the USSR undertook to continue the supply of oil and gas provided the CMEA countries increased their coordination of economic affairs generally and energy systems in particular, developed stronger overall energy efficiency initiatives and off-oil programmes (including the increased use of Soviet gas) and assisted in the modernization of the Soviet economy.

In response to the agreement, the CMEA countries have each developed major programmes intended to maximize the exploration for and development of domestic energy sources. Their commitment to nuclear development reflects the relative availability of alternative sources for the generation of electricity. Thus Bulgaria and Czechoslovakia, relatively short of fossil fuel alternatives, place more dependence on nuclear plans than Rumania or Poland. Hydro-electric opportunities are focused mainly on the Danube where, however, complex international and environmental considerations make progress slow, particularly in the upper reaches. To maximize the regional benefit of new generation facilities, accelerated development of a high-capacity transmission network interconnecting the national systems with that of the USSR is also planned.

Increased efficiency of energy use is another major objective of the member countries. In the long run, structural changes in the industrial sectors are encouraged, including the location of particularly energy-intensive industries in the energy-producing regions of the USSR. In the short term, equipment modification and replacement programmes (e.g. oil refining) and the introduction of energy-efficient operating practices are advocated.

Thirdly, like all oil importers, the Eastern European members of the CMEA seek to reduce their dependency on oil. Fuel-switching programmes from oil to other fuels are to be accelerated by such means as increased output of basic plant equipment (e.g. nuclear reactors in Czechoslovakia), electrification of railroads, and intensification of research and development in such high priority matters as the mitigation of the environmental consequences of the burning of low-grade coals.

Global energy management

Many countries have adopted national energy policies and programmes. Most of these individual countries have been brought together by common energy interests to form international groupings with regional and multi-regional membership. This 'internationalization' of energy management reflects the growth of trends toward interdependence, nationalization and polarization in the global energy scene.

The unprecedented growth of economic activity in the 1950s and 60s was to a large extent energized by oil mobilized by the private sector from the resource base of the countries which were to form OPEC. Thus much of the developed and developing world came to rely on oil imports delivered by the multi-national oil companies. The dependence of the importers on the suppliers developed into a reciprocal relationship as exporting countries came to rely on the income generated from these markets as the main driving force of their modernization plans. Thus there was created a high degree of interdependence between the importers and exporters of energy with the major oil companies as intermediaries.

Because of the crucial importance of the cost and stability of supply of energy to consumers and of the revenue to suppliers, governments became increasingly involved in the management of energy systems. The result was widespread government regulation of and direct participation in all aspects of energy systems with far-reaching results. First, it embedded energy matters into the whole range of political, military and socio-economic aspirations of the nation states of the world. Second, it has redirected a substantial portion of the financial flows associated with the energy industry from the private to the public sector. Third, and as a consequence of the first two results, it significantly changed the character and role of the multi-national energy corporations.

During the long period when the world oil industry was effectively managed by the private sector, commercial market forces, albeit in the context of oligopoly and some political intervention, resulted in a remarkably smooth-flowing and resilient system. In this system such polarization as existed was largely between corporation and government. With the politicization of the energy scene generally and the oil industry particularly, much stronger polarities developed. This first took the form of a rapid escalation of the differences between the objectives of the governments of the producing countries and the corporations operating within their borders. Secondly, after the 1973 oil embargo, strong differences developed between the governments of the major importing and exporting countries and, within them, between those with positive and negative balances of payments. In the 1970s this polarization was accompanied by oil supply interruptions, and a level of uncertainty unknown in previous decades.

As a result of these trends, serious international stresses and strains developed not only in the energy systems of the world but in the entire global economic system. To many observers these warning signs were clear evidence of the need for a concerted move towards global energy management, the first stage of which would be the establishment of an effective dialogue between existing international groups (Ayoub 1979; Foster 1982; Hoffman and Johnson 1981). At this level the key issues were energy supply, price, fuel-transition and efficiency of use.

The energy supply issue had two components both focusing on oil: first, how to avoid, or minimize the effects of, supply interruptions and, second, in the longer term, how to manage the world oil system as it entered the downward side of the depletion curve. The price issue centred on devising means of adjusting to the changing supply–demand scene in a way which produced relatively smooth, incremental price changes within a structure acceptable to the major producers and consumers.

The fuel-transition issue arose from the inevitability of the exhaustion of conventional oil supplies in the long term and the short-term desire for

192

protection from supply interruptions and price escalations. Matters such as the energy sources which should be used to replace conventional oil and the steps necessary to mobilize them needed to be addressed cooperatively. To a varying extent all countries faced the need for fuel-transition but some were in a better position to deal with it than others. The worst off in this respect were the developing countries with a poor energy resource base, negative balance of payments and little technological and organizational experience to draw upon. Finally, the impact of the supply, price and transition issues could be postponed and mitigated if vigorous efforts were made to use energy more efficiently. This required the transformation of energy-using habits formed during an era of plentiful and cheap supply into practices appropriate to a period of increasingly scarce and expensive energy.

Eventually the dialogue required on these issues would have to take place at a truly global scale involving all the groups identified in Figure 7.2. The immediate urgency, however, was seen to be the establishment of an effective

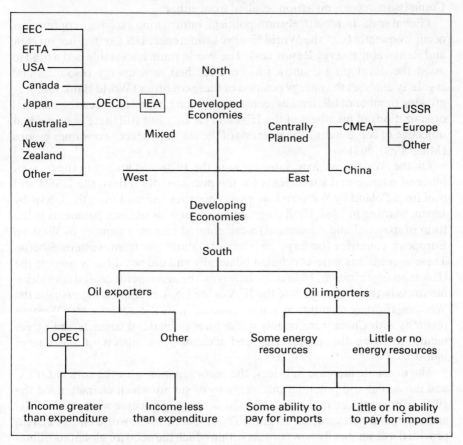

Figure 7.2 The potential participants in a global energy dialogue

and lasting interaction along the 'North–South' axis. In energy terms this would really be a three-way interaction involving two groups of importers, the rich and the poor, and the oil exporters, distinguishing between OPEC and the others. Within each of these groups, however, sub-groups were distinguished from one another on the basis of such criteria as energy resource base, financial position and other affiliations. Such internal differences constitute constraints on the ability of the major groups to maintain cohesion and act in concert

Both the IEA (representing the developed importers) and OPEC (representing the developing exporters) formally identified the promotion of global co-operation as one of their objectives. By 1980 two attempts had been made to create a framework for global co-ordination: the Conference on International Economic Cooperation (CIEC) during 1975-77 and the Independent Commission on International Development Issues (ICIDI), commonly known as the Brandt Commission, in 1978–79 (Maull 1980; Taher 1980). Despite the urgency and effort which characterized the preparation for the final plenary session of the CEIC and the widespread attention accorded to the Brandt Commission report, no action resulted from either.

Global scale, but relatively non-political, information exchange conferences occur frequently (e.g. the World Energy Conference, UN Conference on New and Renewable Energy Resources). The World Bank has established a fund to assist the developing countries to develop their own energy resources and regularly analyses the energy position of these countries (World Bank 1983). A growing number of bilateral agreements are being signed between the national organizations of members of the IEA and OPEC. But still there is no evident objective of coordinating the interests of the major producer-consumer groups (Maull 1980, 303).

On the 'West–East' axis, interaction in the 1970s was mainly in the form of bilateral commercial agreements for the purchase of oil from the USSR and coal from Poland by Western European countries and coal from the USSR by Japan. Starting in 1984, similar agreements (but with advance payments in the form of material and equipment) were entered into by a number of Western European countries for large amounts of natural gas from western Siberia. These agreements were concluded bilaterally and did not directly involve the IEA as an organization. Indirectly, however, the agreements caused considerable stress between members of the IEA as the USA attempted to persuade the Western European countries not to proceed with the agreements. Western relations with China were mainly in the form of bilateral contacts and agreements involving the exploration and technological aspects of the energy industry.

The dramatic increase in prices, the preoccupation with the role of OPEC and the actual and potential interruptions of supply which characterized the 1970s, were replaced in the mid-1980s by decreasing prices, a weakened OPEC and in popular parlance, an 'oil glut'. Thus the state of the world energy system provided a dramatically new context within which the need for global coordination lost much of its earlier short-term urgency. Nevertheless, commentators in the middle of the decade warned of the dangers of complacency for the major problem of energizing Third World countries continues to grow and the 'oil glut' can only be relatively short-lived (Pachauri 1985; Hughes *et al.* 1985). Cowhey argued that both advocates and critics of a 'global social contract' have failed to grasp the complexity of such an arrangement and '. . . underestimated its feasibility and overestimated its benefits' (Cowhey 1985, p. 330). He went on to show how, if international groups of governments are unable to develop effective coordinative strategies, '. . . the rapidly expanding oil corporations of OPEC and the weakened but still influential international oil firms of the OECD could tilt the future of the world oil market.' (Cowhey 1985, Chapter 12).

Summary

This review of energy policy starts with the development of a working framework for the discussion of policy and then applies the framework to a selection of case studies at the national, international and global scale. Public policy is described in terms of four components (perceived issues, formulation, elements and results) set in a context consisting of the state of the system under consideration, its controlling variables (internal and external) and the prevailing ideologies of the political decision makers and the society which they represent.

At the national scale, overviews of the energy policies of three countries with notably different energy systems and socio-political ideologies are presented. Both the USSR and the USA seek to increase their domestic energy production so as to ensure that energy supply does not constrain the evolution of their economies. The USSR, however, relies on hydrocarbon exports for the generation of hard currency and its policy seeks to at least maintain this flow whereas the USA, although a major exporter of coal, imports hydrocarbons. Despite the growth of protectionism in some sectors of its society in the mid-to-late 1980s, its public policy explicitly leaves the source of supplies to be determined by market forces rather than government intervention.

The three countries provide an interesting contrast in the relative importance attached to interventionist and market forces. By the nature of its system, USSR policy objectives are achieved by central government intervention whereas in the USA the Reagan administration (contrary to the previous Carter administration) has reduced intervention and regulation to a minimum. Japan follows a course in which market forces appear to dominate but within the bounds of a carefully prescribed national framework.

OPEC, IEA and CMEA are three international organizations representing exporters, market-orientated and planned importing economies respectively. Such organizations consist of countries between which are many stress-creating differences, but with sufficient community of interest to seek advantages from co-operation and concerted action in proportion to the perceived external threat. The common objective of OPEC members is to maintain and increase their real income from the sale of hydrocarbons in order to sustain the development of their economies and social structure. In the longer term they are concerned to diversify the structure of their economies in part to prepare for the time when their resources are no longer able to sustain large-scale exports.

The chief concern of IEA members, on the other hand, is to protect themselves from administered price escalations and interruptions of supply by means of increasing energy efficiencies and thus reducing total requirements, diversification of supply mix and the areas in which supplies originate and building up short-term reserves in company with an operational plan for allocating and distributing them in an emergency. Eastern European countries have no choice as to their membership in CMEA and have, to a large extent, to fit in with central planning of the USSR upon which they rely for hydrocarbon and nuclear fuel supplies.

The events of the 1960s and 1970s propelled governments first into public policy intervention in the energy industry sector by sector and then into the development of comprehensive, articulated national energy policies. As this increased politicization developed so the constraining role of external, contextual events on what national policies could achieve became clear. International groupings of countries with similar energy characteristics served to internalize these influences to some extent but still left the final level of cooperation, that between the international organizations themselves, unstructured. The crucial questions before the world community are what catalyst will bring about some form of global management of energy involving the major international bodies and the large public and private corporations and when will it occur?

CHAPTER 8

Geography and energy policy: the evolution of the Canadian fossil fuel industry, 1867–1987

The previous chapter dealt with energy policy in a systematic way with selected examples drawn from international groups and the major energy-consuming countries. This concluding chapter deals with only one country, Canada, in order to illustrate in more depth the interaction of the geography of a political unit and the energy policies of its government. In order to search out any persistent characteristics which may exist in the interaction, the study extends over a time period of more than a century.

Even though the focus is upon only one country, and thus is more areally specific than most of the remainder of the book, readers will still find the commentary to be broadly based and general. The data are on a national and provincial basis (rather than individual production and consumption areas), the policy considered is that of the national or federal government (not the provincial governments which have considerable jurisdictional powers over resources) and the long time-period results in the treatment of only the major policy directions at the cost of many short-lived policy developments.

Canada's energy consumption mix has evolved in a manner common to many other economies: from a dominance of wood, to coal, to oil and, by the late 1980s, to a multi-source mix in which oil is still predominant but to a declining extent (Fig 8.1). The spatial distribution of primary energy resources, production and consumption reveals some fundamental and persistent characteristics. First, is the discordance between fuel supply areas (eastern and western Canada) and the major consuming region (central Canada); second is the spatial proximity of US fuel supplies (Appalachian and mid-continent coal and oil fuels) to the major Canadian market area and, the obverse, the accessibility of some US markets to Canadian supplies; third, the spatial intervention of Canadian territory between Alaska and the co-terminus US and of US territory between the Atlantic coast and Montreal; and, fourth, the sharing of rivers and waterways between Canada and the US. Shared features such as the Columbia and Yukon rivers, parts of the St Lawrence and Great Lakes system and the Puget Sound-Straits of Juan de Fuca are relevant in the

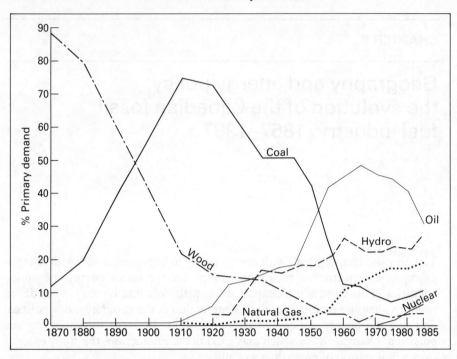

Figure 8.1 Primary energy consumption by source: Canada 1870–1986
Sources: Steward 1978; Canada, EMR 1987

context of hydro-electric generation or transportation of energy commodities or, as in the case of the St Lawrence, both.

The remainder of this chapter treats only the major fossil fuel energy sources: coal, oil and natural gas. For each the evolution of the industries is summarily reviewed and then the relation of geographic characteristics and public policy is examined for selected time-periods. The concluding section draws out the recurring features of this relation and highlights the contribution they make to an understanding of the past, present and, perhaps, future Canadian energy policy.

Coal

Commercial production of coal in Canada increased rapidly in the nineteenth and early twentieth century, levelling off after 1920 but peaking again in 1950 at 19 million tonnes (Fig. 8.2). In the next 15 years output declined to a low of 11 million tons in 1960 after which a rapid increase occurred reaching the all-time high of almost 61 million tons in 1985. During the 120 years between 1865 and 1985, imports matched or exceeded production until 1975 and exports

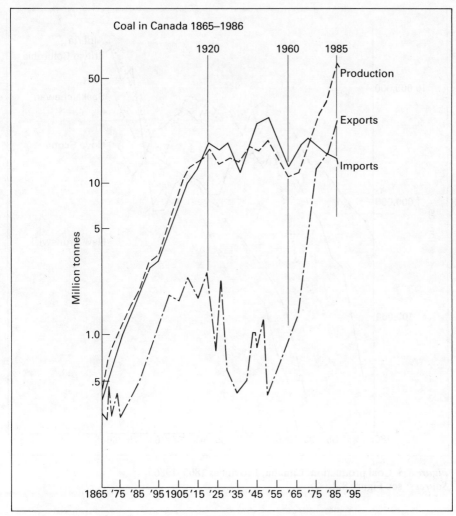

Figure 8.2 Coal production and trade: Canada, 1865–1986
Sources: Urquhart 1965; Carroll 1947; Statistics Canada No. 45–002 various years

ranged from a high of over 75 per cent of production in 1865 to less than 5 per cent in 1955.

Significant commercial production of coal started in the early 1800s in Nova Scotia and New Brunswick, extended to British Columbia before the 1850s and to Alberta and Saskatchewan by the end of the nineteenth century (Fig. 8.3). This broad pattern of peripheral production has persisted throughout the period under consideration. The distribution of coal markets has been equally persistent with the belt from Windsor to Quebec City accounting for more than half of the total consumption of coal (indeed, of all fuels) for the entire period.

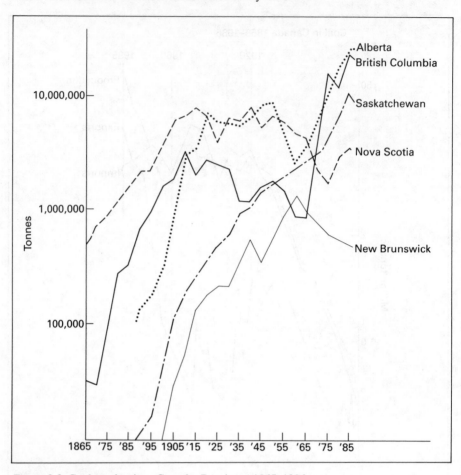

Figure 8.3 Coal production: Canada, Provinces 1865–1986
Sources: see Figure 8.2

This distributional pattern of production and consumption sets the scene for spatial interaction between the periphery and the core, of coal moving from the eastern and western producing areas to the heartland of central Canada. However, as Figure 8.4 illustrates the effective development of this interaction has been thwarted by an intervening opportunity: the coal resources of the northeast US. Canadian coal has never been able to capture even as much as 10 per cent of the Ontario market. Until World War I, coal from Nova Scotia and the United Kingdom supplied different sectors of the Quebec market. After 1920, US coal replaced that from the UK and provided increased competition for Nova Scotian production. Coal from Alberta and, to a lesser extent, British Columbia and Saskatchewan, penetrated the Winnipeg and Ontario lakehead market during World War I but, despite considerable efforts, did not reach into southern Ontario in significant quantities until the 1970s.

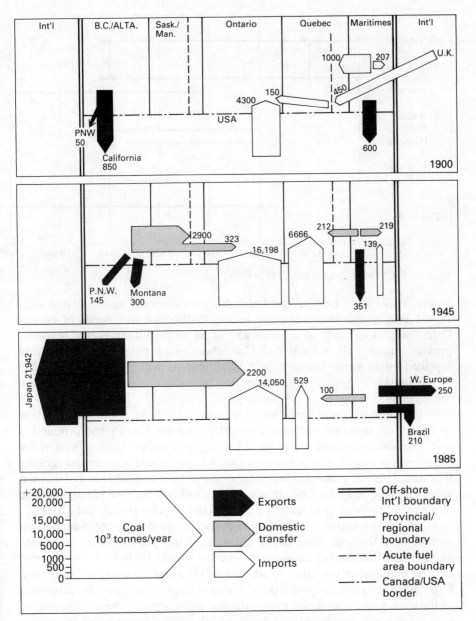

Figure 8.4 Coal trade and transfer; Canada, 1900, 1945, 1985
Sources: see Figure 8.2

Coal has been exported from both Nova Scotia and British Columbia from the beginning of commercial production in those areas. From the former, coal was shipped to the coast cities of the northeast US until competition from US coal and, later, oil, effectively closed the market. Small scale export oppor-

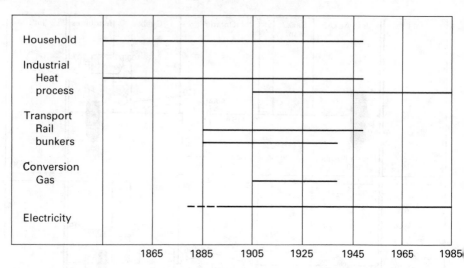

Figure 8.5 Canada: uses for coal, 1865–1985

tunities to Western Europe and Latin America emerged intermittently after World War I and more regularly in the 1980s. The fuel needs of coastal Californian cities and, to a lesser extent of more distant Pacific markets, provided major outlets for British Columbian coal for almost a century until replaced by oil in the 1920s. After trial shipments in the 1950s, a large-scale export market to Japan began to develop reaching a maximum of 27 million tonnes in 1985.

Four major groups of variables have operated to bring about the evolution of the spatial structure of the coal industry: end-use, quality and production cost, transportation cost and public and corporate policy. During most of the nineteenth century the major use for coal was household heating and cooking followed, later, by general industrial use, smelting, transportation and thermal electricity (Fig. 8.5). By 1985, the only large scale uses of coal remaining were in the metal smelting (particularly iron) and the thermal electric industries.

Most of these end-uses require coal of specific quality for optimum performance. Thus the household market showed a strong preference for anthracite and the smelting market for bituminous coking coal (Table 8.1). The thermal electric market, on the other hand, can make use of a variety of coals from bituminous to lignite by matching the design of equipment to the characteristics of the coal. Once committed to a particular coal supply, however, changes in coal quality will usually require more or less capital intensive alterations in order to achieve the initial design performance. The need to match coal quality and end-use has significant spatial implications for the industry. For example, except for the deposits in north-western British Columbia (known since at least the beginning of the twentieth century), Canadian coal resources do not include significant anthracite deposits. Thus during the long period when anthracite was the preferred household fuel, supplies had to be imported from South

202

Table 8.1 Canada: coal classes, use and quality indicators

Class	Major use	Heat content BTU/lb	Ash/sulphur
Bituminous			
Coking	Smelting	13,000–14,000	Varied – sulphur
Thermal	Transportation		generally high in
	Industry, heat		E. N. America
	Electricity		and low in west
Sub-bituminous	Household	8,500–11,500	As for bituminous
	Electricity		
Lignite	Electricity	6,500–8,500	High ash

Wales or Pennsylvania. As will be seen below, public policy initiatives were introduced in an attempt to develop substitutes.

Coal production costs are considerably influenced by the conditions of occurrence as well as the quality of coal produced (Chap. 4). In general, production costs have been higher in eastern Canada than in the western provinces. As large open-pit mines were developed in British Columbia, Alberta and Saskatchewan and the Nova Scotia mines were having to go deeper and further out under the sea this differential has been accentuated.

The ability of Canadian coal to penetrate the central Canadian markets is not only a function of quality and production costs but also of transport costs. In general terms it is almost three times as costly to transport coal by rail as by water. Consequently, the longer the overland distance to market the greater the transport costs. Thus western Canadian coal is at a considerable disadvantage compared with US coal in the central Canadian market. Furthermore, in most of Canada, there is a marked seasonality in the operating conditions of transport systems, ranging from frozen waterways to heavy snow, which posed severe marketing problems in the earlier years of the period under review. Secondly, for both the rail and water movement of coal, availability of 'back-haul' loads could significantly reduce transport costs. Unfortunately, in most cases such freight is not available in Canada.

Policy considerations constitute the fourth set of variables helping to shape the spatial structure of the coal industry. For complete coverage it would be necessary to consider both public (to include Federal, Provincial and even foreign governments) and private policy (to include coal producers, users and transporters). Here the emphasis will be on public federal policy. A conceptual framework within which policy characteristics may be identified and analysed has been presented in Chapter 7. This framework will be generally applied in the following section on coal and the other energy sources dealt with in this chapter.

Federal policy

For purposes of policy analysis it is convenient and, to a considerable extent, empirically sound, to divide the evolution of the coal industry in Canada into three major periods (1867–1920, 1921–60, and 1961–85) and a short, most recent period (1986 onwards). In each of the major periods it is also helpful to introduce sub-divisions in order to highlight times of greater (or lesser) public policy activity. Some of the most notable characteristics of each of the major periods are summarized in Table 8.2 and the text that follows aims to elaborate upon this condensed material.

1867–1920: The emergence of the coal industry In 1867 the spirit of Confederation included hopes of Canadian unity, a broadened base for a Canadian economy and the associated development of inter-provincial trade. In this context it is not surprising that Nova Scotia had high expectations of expanding its coal industry by means of shipping to both the central Canadian market (already supplied from the US) and to New England. Early debates (1870–76) in the House of Commons on Canadian fuel policy were led by C. Tupper and T. McKay from Nova Scotia (McDougall 1982) who argued for the imposition of a tariff on US coal imports in order to make coal from their province competitive in the major Canadian market. They made the case in terms of Canadian unity, benefits to the Nova Scotian (and, thus, they argued, Canadian) economy and, to a lesser extent, dependability and independence from foreign supplies. Ontario representatives were not enthusiastic and raised questions about how far the people of Ontario should make sacrifices in price and quality in the name of Canadian unity.

Lengthy and frequent debates on this issue continued throughout the period. However, the only real policy decision by successive governments was the application of a tariff on US (not British) imports. Presumably in an attempt to balance the regional interests of the producing and consuming provinces, the tariffs were relatively small (50–60–75c/tonne) and were not sufficient to either diminish the flow of US coal into Ontario or increase significantly the amount of Nova Scotia coal moving beyond Montreal.

The economic activity associated with World War I led to expanded markets for coal in both central Canada and Nova Scotia. However, the marketing of Nova Scotia coal up the St Lawrence was severely limited by shipping problems. Furthermore, by 1917, the US industry was no longer able to supply Winnipeg and north-western Ontario, allowing western Canadian producers into this market. In fact it was only with great effort that US supplies had been able to meet the needs of south-western Ontario and Quebec, and by the end of hostilities central Canadians had come to realize the risks associated with this dependence on imports. The wartime fuel controller, C. A. Magrath, wrote in his final report that '. . . Canada is far too cold a country to depend on [Pennsylvania] alone', and that:

The coal operators, both in the East and the West, should be encouraged as far as possible to make the most of the domestic markets and thereby decrease the amount of coal imported. (Magrath 1919, pp. 37 and 69).

1921-60: Stagnation and decline The matter of 'National Fuel supply' became a major issue in the reconstruction of the Canadian economy after World War I. The rhetoric referred repeatedly to the 'acute fuel problem' and in April 1921 a Special Committee of the House of Commons was established to '. . . enquire into the fuel supply of Canada . . .' (Canada, House of Commons 1921, p. 111). The committee recommended the appointment of '. . . an officer of Government [who should] keep in close touch with the fuel situation of Canada' and advocated the development of other sources of energy (hydro and peat), the reduction of the cost of water transport, and the encouragement of the use of domestic coal rather than imported anthracite. In response to these recommendations, the only action taken by the government was to establish the Dominion Fuel (later, Coal) Board.

The debate over the supply of fuel to what was called the 'Acute Fuel Area' (Winnipeg–Montreal) became considerably more urgent as the result of the dramatic impact of the US coal strike in 1922–23. As a consequence the government referred '. . . the whole question of fuel supply for Canada . . . and [the means] whereby Canada may become self-sustaining as regards fuel . . .' to the Select Standing Committee on Mines and Minerals (Canada, House of Commons 1923, p. 3). Also in 1923, a Special Committee of Senate undertook an examination of the same topic.

The hearings before these committees and the debates in the legislative bodies were characterized by a remarkable agreement between spokesmen from both producing and consuming provinces. There was almost a 'national' consensus focusing upon the reduction of the cost of transporting coal from the periphery to central Canada, the prospect of substituting Canadian coal (particularly western coal) for imported anthracite from the US and the United Kingdom, and the use of peat.

In response, in 1928, the Federal government instituted the payment of subventions to assist in the transport of coal from both Nova Scotia and Alberta to the 'Acute Fuel Area' and, in 1931, for bunkering in British Columbia. In 1927 and 1930 respectively, the Domestic Fuel Act and the Canadian Coal Equalization Act became law. The former provided for Federal assistance in the construction cost of plants gasifying Canadian coal and selling the coke in the household market as a substitute for imported anthracite. The latter provided for subsidies to encourage the use of Canadian coal in the iron and steel industry. Notwithstanding these policy initiatives and in the face of the depression, the coal industry was able to do little more than maintain its output during the 1930s and imports showed no long-term decrease.

World War II and Canada's role as a major supplier of materials, munitions and weapons brought about a considerable increase in economic activity and an associated demand for fuel. Production, imports and transfer of coal all increased but as the end of the war drew near, the future energy supply of

Table 8.2 Canadian fuel policy: coal, 1867–1986

	1867–1920 Emergence	1921–1960 Stability and decline	1961–1985 Revival	1986 Incipient decline
Context	– Rapid rise in use of coal	– Rapid rise in use of oil	– Oil dependence peaked	– Expectations of rising off-shore market for thermal coal
	– Canadian Confederation			
	– Large supply from Penn., W. Virginia, Ohio	– Decline of Penn.: Anthracite US coal strike, 1922–23		– Questions about sulphur content
			– Emergence of Japanese market for metallurgical coal	– Decline of Japanese market
	– Regional markets in N.E. and West coasts of US	– Markets lost to oil and natural gas	– Potential Pacific Basin market for thermal coal	
Perceived issues	Producing Provinces: Increase production	Maintain economic base →		
	Consuming Provinces: Maintain reliable, least cost and acceptable grade supply		Reliability of supply Concern for regional disparities →	Balance of payments Regional economic development
	Federal: National unity and independence			Quality of environment

Objectives and targets

Producing Provinces:
Increase own-use, exports, inter-provincial trade and decrease imports

Consuming Provinces:
Increase access to reliable, low-cost, suitable grade supplies

Federal: Develop inter-provincial trade, reduce imports

Strategies and means

Federal: Tariffs

Development of domestic markets

Coke industry

Subventions and grants
Transportation
Exports and bunkers

Devco

Ontario Hydro
Sysco

Task Forces and Working Committees

– Parliament & Parliamentary Committees

– Royal Commissions: 1925, 1947, 1960

– Fuel Controller, 1915–1919

– Dominion Fuel Board 1922   1970
– Domestic Fuel Act 1927–1932
– Can. Coal. Eq. Act 1930   1970
– Coal Prod. Assis. Act   1970

Canada once again became a concern. In October, 1944, C. D. Howe, then Minister of Munitions and Supply, persuaded the government to establish a Royal Commission to undertake a full enquiry into the coal industry in Canada.

The Commission reported in 1947 and recognized that:

> Independence may be physically possible, but it is too impractical to merit further attention . . . this view does not preclude the movement of some Canadian coal into the [central Canadian] market with assistance (Canada, Royal Commission 1947, p. 582).

The report supported continued assistance in the name of 'fairness', the economic dependence of producing provinces (particularly Nova Scotia) upon coal, the value of a national dispersion of industry and in being prepared for emergency situations. The Commission advocated a combination of tariffs and transport subventions (not production subsidies) as a '. . . permanent government policy' but, in respect of Nova Scotia, did not support either marketing assistance or nationalization. The Commission also recommended the establishment of a Statutory Board with a full-time Chairman to '. . . keep Canada's energy requirements under continuous review and to advise upon and administer transportation subventions' (Canada, Royal Commission 1947, p. 583). The only direct government response to the report was the Coal Production Assistance Act of 1949 which authorized the payment of low interest loans for the purpose of modernizing and mechanizing Canadian mines.

Despite these measures the production of coal declined in all provinces during the 1950s in the face of intense competition from oil and natural gas. By 1955, oil products had become the major fuels in the Canadian economy and only the integrated iron and steel and thermal electricity industries remained as markets for coal. Furthermore, the discovery of oil in Alberta in 1947 and subsequent hydrocarbon discoveries in the western sedimentary basin led to a government pre-occupation with the marketing of oil and natural gas and the relative eclipse of concern for coal. Nevertheless, as a consequence of the 1958 report of the Royal Commission on Energy (which, despite its title, focussed exclusively on hydrocarbons), the Federal government established a National Energy Board. Although both at its inception and subsequently this Board had little concern with coal, its terms of reference suggest a link with recommendations of the 1947 Commission on Coal to appoint a Statutory Board.

This period of stagnation and decline ended with one more enquiry into coal. In 1959 a Royal Commission on Coal was established to make recommendations concerning costs of production, marketing and 'reasonable government measures' (Canada, Royal Commission 1960). The recommendations focused on two classes of subsidy, 'Basic' and 'Social', to be applied differentially to coal from Nova Scotia and the West, and advocated the removal of export subsidies and the thorough study of ways of improving efficiency. No action resulted from the Commission's report.

1961-85: Revival of the industry Although the years 1961 to 1985 were dominated by concerns for oil and gas, world events developed in a way which was to have a major influence on the coal industry of western Canada. The rapid expansion of the Japanese iron and steel industry led to a strong demand for supplies of metallurgical coal. Trial shipments from British Columbia and Alberta as early as 1953 had been well received and by the early 1960s, aided by federal subventions, exports exceeded one million tonnes. Federal funds were also provided to aid in the construction of the large-scale transportation and handling facilities that were required and, by 1985, shipments exceeded 25 million tonnes. Furthermore, in response to concerns about US supplies, Ontario Hydro entered into an agreement with western coal suppliers for up to three million tonnes per annum, delivery of which commenced with the completion of new coal-handling facilities at Thunder Bay in 1977.

The industry in Nova Scotia continued to face difficulties. Ever increasing production costs of coal with relatively high sulphur and ash content made export and transfer markets difficult to enter. Declining production through the 1960s and early 1970s brought locally severe economic and social disruption in a region already hard hit by a general decline in economic activity. In response to these problems, although tariffs were removed in 1967 and transportation subventions in 1968, successive federal governments made large-scale, direct investments in both the coal-using and coal-production industry of Nova Scotia via the Nova Scotia Development Corporation and, most recently, the Sydney Steel Company. As a consequence of these regional developments and small export markets in western Europe and Latin America, production in 1985 reached the levels of the beginning of the century.

By the early 1980s, ominous signals were coming from the Pacific Basin coal buyers. In the face of a global economic downturn and a re-structuring of the economy, Japan's metallurgical coal requirements started to decrease. This, in conjunction with the large productive capacity of existing and new global coal supplies, led also to declining prices and increasing difficulties for the export-oriented producers of Alberta and British Columbia.

Post-1985: Incipient decline The weakening of the Pacific metallurgical coal market continued in the second half of the 1980s and the western Canadian coal industry is facing a period of marketing constraints. Major producing corporations in British Columbia are already having serious financial difficulties, and, on the east coast, the modest revival of the Nova Scotia industry remains fragile. Even the decision to re-equip the Sydney Steel Company (declared, in 1987, to be the sole supplier of rail to the CNR) may not be sufficient to sustain it.

With the incipient decline of production in both major coal areas, federal assistance is once again being sought by the producing regions. Reminiscent of the 1920s and early 30s, in 1984 the governments of Alberta and Ontario formed a task-force of experts to examine the prospect of increasing the use of western coal in Ontario. In October of the same year a meeting of the provincial

Ministers of Environment recommended the establishment of a federal-provincial 'Task Force on Expanded Use of Low-Sulphur Western Canadian Coal.' The group was duly appointed and reported in June, 1986.

The motivation for the formation of the task-force came from two directions: first, the potential decline of export markets for western coal and the resulting surplus production capacity (a familiar theme) and, second, an environmental quality problem in Ontario (a new issue). The Ontario government has enacted legislation which requires the reduction of sulphur dioxide emissions into the atmosphere. The impact of this legislation upon the market for coal in Ontario is mainly felt by its effect upon the electricity supply and iron and steel industry. These industries must modify existing installations to meet the new standards and introduce conforming options for future expansion.

For both existing and future installations, one answer (out of several) would be to switch from relatively high-sulphur US coal imports to low-sulphur western Canadian transfers. In its report, the Federal-Provincial Task Force indicates that such a shift is technically feasible and would confer significant economic benefits to Alberta and British Columbia but at the expense of higher costs to Ontario users (Canada, Federal-Provincial Task Force 1986). In response to federal government initiatives for active consideration of the report, regional interests have expressed entirely predictable views: enthusiastic support from Alberta and British Columbia and conditional interest from Ontario. To date, neither transportation nor use subventions have been explicitly mentioned. Rather, emphasis has been placed upon reducing production and transportation costs with a hint of federal assistance in converting existing Ontario plants to the use of western coal. In mid-1987 the federal government established an 'action committee' consisting of both political and technical personnel from Ottawa and the involved provinces to study the issue.

Oil and natural gas

Small-scale production of hydrocarbons commenced in Canada in the nineteenth century with oil and gas in south-western Ontario and gas in New Brunswick and Alberta. However, only in Alberta did this early production foreshadow large scale resources. These were discovered on a modest scale in 1914 and a large scale in the two decades between 1947 and 1967 (Fig. 8.6). Since 1970, and despite extensive exploration (Fig. 8.7), additions to the light oil resource base have not kept pace with production. In the 1980s, most discoveries have been in the frontier areas in northern Canada and off the Atlantic coast. The discovery and production costs of this oil are high as will be the transportation cost to markets. Intensive delineation and development drilling in the heavy oil and bitumen deposits of Alberta and Saskatchewan have confirmed the existence of a large conditional reserve base. Although production from these resources is increasing, major expansion requires still

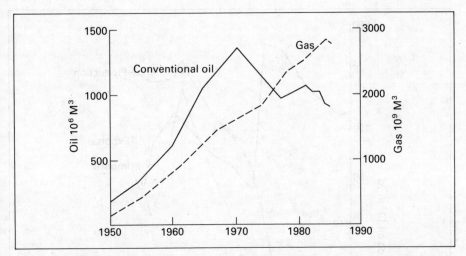

Figure 8.6 Established reserves of conventional oil and natural gas: Canada, 1950–86
Source: *Oilweek* various issues

further developments in production technology and a minimum price of approximately $20/barrel.

In the early days of the Canadian hydrocarbon industry most natural gas was discovered in association with oil and, in the absence of major distribution systems, the majority of the produced gas was flared. By the 1950s, the value of the growing volume of associated gas reserves was recognized and non-associated gas discoveries became increasingly sought after and discovered reserves of gas increased steadily until 1984 (Fig. 8.6).

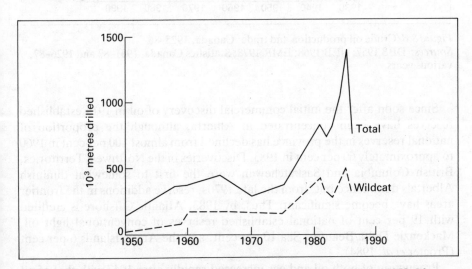

Figure 8.7 Canadian hydrocarbon exploration: total and wildcat drilling, 1950–86
Source: *Oilweek* various issues

211

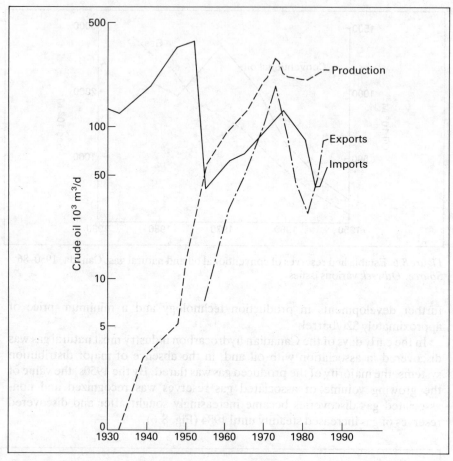

Figure 8.8 Crude oil production and trade: Canada, 1925–86
Sources: DBS 1957; NEB 1968; EMR 1978; Statistics Canada, 1931–87 and 1926–87, various years

Since soon after the initial commercial discovery of oil in 1914 established reserves have been concentrated in Alberta, although the proportion of national reserves in the province has declined from almost 100 per cent in 1960 to approximately 60 per cent in 1983. Discoveries in the Northwest Territories, British Columbia and Saskatchewan were the first to somewhat diminish Alberta's dominance but, from the late 1970s, reserve additions in the frontier areas have become significant. Thus, by 1983, Atlantic Offshore is credited with 19 per cent of national established reserves of conventional light oil, Mackenzie Delta/Beaufort Sea 10 per cent, and the Arctic Islands 6 per cent (Procter *et al*. 1984).

Production of both oil and gas increased rapidly after 1947 with that of oil peaking in 1973 and gas in 1985 (Figs 8.8 and 8.9). The decline in oil production

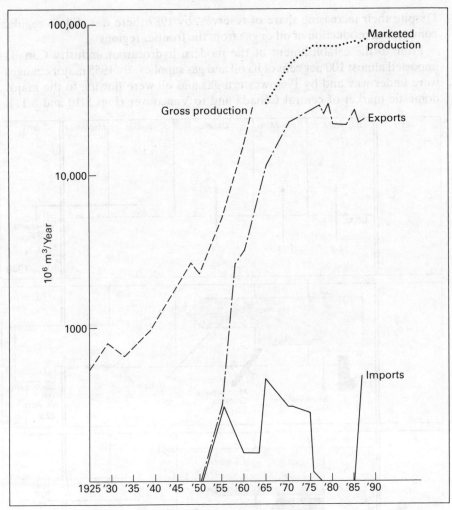

Figure 8.9 Natural gas production and trade: Canada, 1926–86
Sources: Simpson *et al.* 1961, 1964; Canada, Department of Energy, Mines and Resources, 1978 and NEB 1986

between 1973 and 1980 reflects the diminishing conventional light oil resource base while the subsequent partial recovery to 1986 is largely due to increased output of heavy oil and synthetic crude from bitumen. The failure of gas production to increase significantly after 1975 is not related to supply constraints but rather to export marketing problems in the United States.

Although Alberta's dominance in the distribution of reserves has been declining, the province remains the centre of hydrocarbon production with over 80 per cent of national oil output (followed by Saskatchewan with 13 per cent) and 90 per cent of gas (followed by British Columbia with 8 per cent).

213

Despite their increasing share of reserves, by 1987 there was still no regular commercial production of oil or gas from the frontier regions.

Prior to the establishment of the modern hydrocarbon industry Canada imported almost 100 per cent of its oil and gas supplies. By 1955 major changes were under way and by 1960 western gas and oil were flowing to the major domestic market of central Canada and to Vancouver (Figs 8.10 and 8.11).

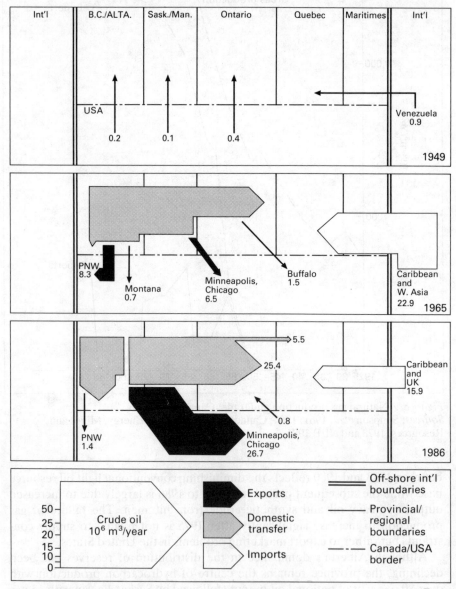

Figure 8.10 Crude oil trade and transfer: Canada, 1949, 1965, 1986
Sources: see Figure 8.8

Concurrently, exports to the United States were developed with oil and gas moving to the north-west and upper mid-west states and, by the late 1960s, gas to California. Between 1955 and 1965 approximately one-third of Canadian oil production was exported to the United States, and by 1975 as much as one-half of the output flowed over the border. In the early 1980s, the proportion exported dropped to less than 20 per cent only to rise again to 34 per cent in

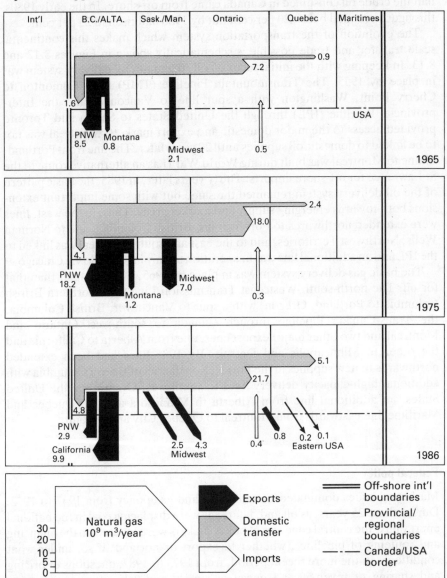

Figure 8.11 Natural gas trade and transfer: Canada, 1965, 1975, 1986
Source: See Figure 8.9

1986. Exports of gas quickly rose to approximately 35 per cent of production after 1955, climbed to almost 50 per cent in the late-1960s, then declined to less than 30 per cent of production in 1986 reflecting decreased sales to the USA. While the trade in natural gas has been essentially export-dominated with only very small and irregular imports, that of oil has been characterized by continued imports. Indeed, imports of oil exceeded exports up to 1982 and in 1975 over half the crude oil consumed in Canada came from off-shore. In the early 1980s, this figure dropped below 20 per cent but by 1986 had risen to 25 per cent.

The evolution of the transportation system which makes this continental scale transfer and trade possible is schematically shown in Figures 8.12 and 8.13. In keeping with the initial focus on oil, the major oil delivery system was in place by 1955. The Transmountain Pipeline (TMP) from Edmonton to Cherry Point, Washington with a spur line to Vancouver and the Inter-provincial Pipeline (IPL) through the United States to Sarnia and Toronto provided access to the major domestic and export markets. Montreal was not to be linked to domestic oil supplies until 20 years later. The line from Portland, Maine to Montreal was built during World War II as an alternative route to the St Lawrence for off-shore imports. Thirty years later, in 1985, the basic pattern of the oil delivery system remained the same but with some important extensions both towards emerging supply and market areas. Thus, in the west, lines were extended northwards to Fort St John, British Columbia, and to Norman Wells, Northwest Territories, and to the east and south Montreal was linked to the IPL and an additional line was in operation to Sarnia, Ontario via Chicago.

The basic gas delivery system was in place by 1965, a decade later than that for oil. The north-south Westcoast Transmission Line from northern British Columbia to Portland, Oregon, with a spur to Vancouver, British Columbia, the Trans-Canada Pipeline through Canada to south-west Ontario and Montreal and two other major export lines, one from Alberta to California and the other to Minneapolis and Detroit. By 1985 lines had been extended northwards to new supplies in the Fort Nelson district of British Columbia with additional high-capacity delivery lines to south-west Ontario via the United States, an additional line from Alberta to Minnesota and the Quebec and Maritime line eastwards from Montreal to Quebec City and beyond.

Federal policy

Marketing issues dominated Canadian oil and gas policy from 1947 to 1972. During these 25 years, as oil and, later, gas were displacing coal in the national energy mix, the central concerns of policy makers were focused on the financing and routeing of pipelines, whether to export or not and, if so, under what conditions. In the more recent period, from 1973 to 1987, questions of pricing and sharing of revenues, adequacy of supply and self-sufficiency, foreign control and federal-provincial-industry relations were added to the list. The basis of these perceived issues and the objectives and strategies adopted by

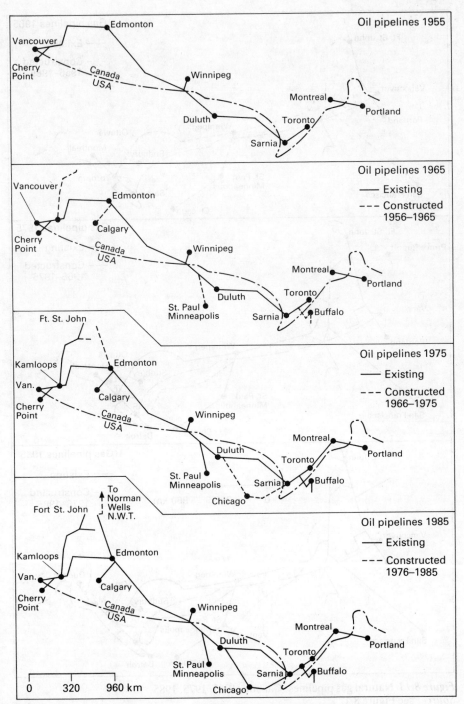

Figure 8.12 Crude oil pipelines: Canada, 1955, 1965, 1975, 1985
Sources: Canadian Imperial Bank of Commerce various years; *Oilweek* 1986a

The evolution of the Canadian fossil fuel industry 1867–1987

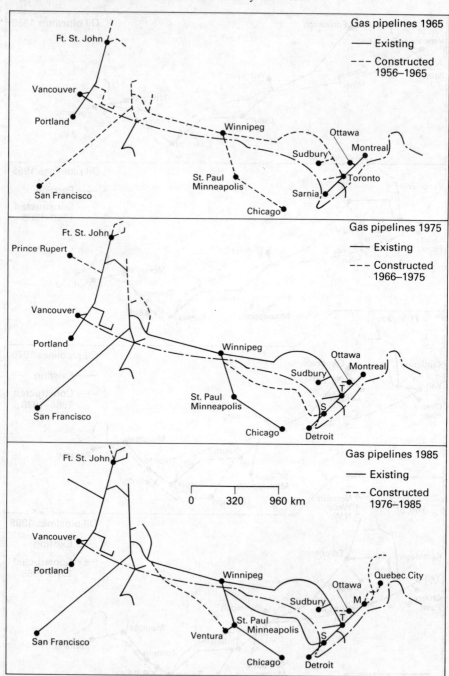

Figure 8.13 Natural gas pipelines: Canada, 1965, 1975, 1985
Source: see Figure 8.12

successive federal governments between 1947 and 1987 are considered below in five relatively distinct periods and summarized in Tables 8.3 and 8.4.

1947-61: The establishment of the oil and gas industry   Throughout this period oil and gas reserves grew rapidly. Industry investors were anxious to market these reserves and thus secure a return on their investment which could be used in part for continued exploration. Governments of the resource provinces saw great opportunities to augment their income and develop downstream oil and gas processing. Consumer provinces watched with envy and supported transportation proposals which promised to enhance both their energy supply options and their economic development. The federal governments of the period (Liberal to 1956, Conservative thereafter) sought to establish a fiscal and regulatory context within which the growth aspirations of the producers and the least cost–greatest dependability objectives of consumers could be realized. For the most part this meant juggling the conflicting objectives of those who espoused nationalism or economic efficiency. The former supported Canadian fuels for Canadian markets, coast-to-coast and all Canadian routes for the necessary transportation systems. The economic efficiency, or free-marketeers, advocated seeking the best return on investment and accepting the greater dependence on United States routes and markets that this would entail.

The first Federal initiative was the Pipe Lines Act, 1952 which required that all companies proposing to move oil or gas out of a province must be incorporated by an act of Parliament and then have the proposal approved by the Board of Transport Commissioners (BTC). In practice the Board also required the companies to have the approval of the government of the producing province to remove the oil or gas before their application would be considered.

The parliamentary debates on the Pipe Lines Act, the acts of incorporation, and the hearings before the BTC raised many of the issues which arose during the evolution of the coal policy. It was repeatedly argued by producers that export markets should be energetically sought, and by the federal government and consumers that the least-cost transportation systems should be developed. In order to achieve least-cost status such systems would have to be large-scale which, in turn, meant that they would have to carry large volumes often only available in United States markets. In opposition to these views, the 'nationalists' strongly voiced the desirability of all-Canadian routes for pipelines in order to maximize the benefits to communities along the route and to remove any chances of United States actions affecting supply. Consumers stressed the importance of dedicating established oil and gas reserves to current and future Canadian requirements and only permitting exports after these requirements had been provided for. Otherwise, they argued, established, less expensive reserves would be exported leaving future Canadian needs to be met from undiscovered, more expensive sources.

In 1953, the federal government issued a policy statement which stated that oil should move '... from the source of production to refineries within

219

Table 8.3 Canadian fuel policy: oil and gas, 1947–78

		Establishment of industry 1947–61	Consolidation and regulation 1962–72	Politicization and interaction 1973–78
Context	Global	– Rapidly rising hydrocarbon reserves – Low and stable oil price – Rapidly rising oil consumption	– Increasing strength of OPEC – Increasing world supplies and decreasing *real* price	– 1973 Start of rapid increase of real world oil price
	USA	– Became oil importer – 1959 Restrictions on imports, Canada excluded	– Imports increasing – 1969 Discovery of oil and gas, Prudhoe Bay	– Search for modes and routes by which to market Prudhoe Bay oil in USA
	Canada	– Rapid shift from coal to oil – 1947 Discovery of Leduc oilfield. Thereafter increasing reserves of oil and gas – Canadian oil *more* expensive in central Canada than imported	– Increasing exploration in Canadian Arctic and offshore East coast	– Confirmation that reserves of light oil peaked in W. Sedimentary basin; frontier oil discoveries disappointing, gas more encouraging – Canadian oil *less* expensive in central Canada than imported
Perceived issues	Supply/markets	– Marketing of hydrocarbon discoveries – Domestic market/export market – All Canadian versus through US	– Size of conventional oil reserves – Determination of 'surplus for export' – Northern pipeline proposals	– Cost and reliability of off-shore supplies; adequacy of domestic supplies – Marketing gas from Canadian Arctic – Transport of Alaskan oil/gas through Canada – Impact of world price increases on Canada
	Price			– Division of increased revenues between Federal and Provincial governments and industry
	Revenues		– Reform of tax structure favourable to oil and gas industry – Public reliance on industry-generated data – Canadian participation in and regulation of Northern developments	– Division of increased revenues between Federal and Provincial governments and industry
	Jurisdiction and regulation	– Control of industry by foreign companies – Selection of pipeline routes and regulation of inter-Provincial transfer and exports		– Federal versus Provincial jurisdiction – Continued absence of major Canadian hydrocarbon company

220

Section	Category	Column 1	Column 2	Column 3
Objectives	Supply/markets	– Access markets which provide highest returns – Regulate exports to protect Canadian needs	– Expansion of production and markets	– Self-sufficiency ⟶ self reliance – Increase alternative oil supplies (e.g. tar sands) – Discontinue exports
	Routeing	– Establish control of pipeline development – Develop *least cost* routes to markets – Determined by market forces	– Support routes favouring exports	– Extend oil pipeline eastwards – Protect Canadian economy from world price increase
	Price		– Regulate export prices	
	Revenues	– Structure taxes to encourage industry	– Review of changes necessary to increase government revenue	– Increase Federal revenue and dampen growth of Provincial and corporate revenues
	Jurisdiction and regulation	– Establish appropriate regulatory and administrative agencies	– Establish additional public agencies and increase role of existing – Depoliticization of decision making	– Limit producer provinces' power with respect to pricing and production
Programmes and means	Supply/markets	– 1951 Pipelines Act requiring export permit for oil and gas – 1957–1959 Royal Commission on Energy	– 1970 Public financing in Panarctic – 1972 NEB reports conventional oil reserves insufficient to sustain export levels	– 1973 Restrictions on exports – 1974 Joint Federal/Provincial industry funding of oil sands – 1975 Conservation, renewable energy, and R & D programmes – 1976 Interprovincial pipeline extended to Montreal
	Routeing	– 1961 National oil policy		
	Price			– 1973 Alberta strengthens control over pricing and marketing – 1975 Petroleum Administration Act establishes Federal price control
	Revenues			– 1973 Introduction of oil export charge and oil import compensation programme
	Jurisdiction and regulation	– 1959 Establishment of National Energy Board (NEB) – 1961 Canada Oil and Gas Lands (COGL) regulations limiting *production* licences to Canadian-controlled companies	– 1966 Establishment of Department of Energy, Mines and Resources (EMR) – 1968 Task force on Northern development – 1972 Publication of 'Guidelines for Northern development'	– 1975 Formation of National Oil Company (PETROCANADA)

Table 8.4 Canadian national energy policy: 1979–87

		Comprehensive intervention – The National Energy Program, 1979–83	Disengagement and deregulation, 1984–87
Context	Global	– 1979 Dramatic OPEC price increase followed by stable, then declining prices – Recession	– Rapid decline in world prices early 1986, followed by slow rise – Diminished role and instability of OPEC – Growth of protectionism
	USA	– Overall decline of energy demand but continued increase in oil import: deregulation of gas industry and emergence of 'gas bubble'	
	Canada	– Canadian oil *lower* price than imported in central Canadian market – Decreasing conventional oil reserves in Western Sedimentary basin; disappointing frontier results; gas reserves increasing – Liberal government, 1980–84	– Canadian oil *higher* price than imported in central Canadian market – Large gas reserves; declining conventional oil in Alberta – Conservative government, September 1984
Perceived issues	Supply/markets	– Increasing dependence on oil, increased imports; concern for cost and security of oil supply; continued search for markets for gas	– Highly regulated supply and marketing conditions
	Routeing	– Need for coast to coast oil and gas delivery systems	– Suitability of existing pipeline system in relation to developing resource and market patterns
	Price Revenues	– Set by international forces rather than national – Insufficient revenues to national government; too large to producing provinces and corporations	– Degree of regulation – Tax burden too high
	Jurisdiction and regulation	– Industry controlled by foreign corporations – Degree of control over hydrocarbon resources exercised by Alberta	– Poor relations between Federal government and Provincial governments and corporations – Degree of Federal intervention

Objectives	Supply/markets	– Reduce consumption of oil, discontinue imports by 1990 – Increase supply of domestic oil, increase overall efficiency of use	– Develop energy resources as an engine of economic growth
	Routeing	– Enlarge existing oil pipeline to Montreal; extend gas line to Quebec city and Maritime provinces	
	Price	– Establish 'Made in Canada' price structure below world price but rising toward it, increase gas price but keep below oil	– Develop system of taxes and incentives which provides fair treatment of consumers and producers, stimulate investment in new resources and allow market sensitive mechanisms to operate – Adjust taxation structure to maintain industry in face of low prices
	Revenues	– Establish 'fair' distribution between Federal government, government of producing provinces and corporations	– Cooperation between Federal and Provincial governments and industry
	Jurisdiction and regulation	– Assert 'National interest' while recognizing Provincial jurisdiction over resources – Increase Canadian ownership of hydrocarbon industry to 50% by 1990	– Recognition of Provincial rights to off-shore hydrocarbon resources
Programmes and means	Supply/markets	– Targets for decreasing oil consumption, conservation and efficiency targets and programmes – Petroleum Incentive Program (PIP) to sustain frontier exploration	– Rapid removal of most supply and marketing regulations so as to permit operation of 'market mechanisms'
	Routeing	– Fiscal incentives for enhanced oil recovery, heavy oil and tar sands	– Flexibility toward product mix (e.g. light/heavy, NGL, products) carried in oil pipelines
	Price	– Federally determined national price schedule increasing regularly to world price; distinction between 'new-old', 'light-heavy' oil	– Deregulation of oil and natural gas prices
	Revenues	– Petroleum compensation charge – Petroleum ownership special charge – Canadian ownership special charge – Export tax	– Most NEP taxes removed by 'Western Accord' and 'Agreement on Natural Gas Markets and Prices'
	Jurisdiction and regulation	– PETROCAN authorized and funded to acquire foreign companies operating in Canada – Exploration and production from 'Canada Lands' limited to Canadian companies – Petroleum Monitoring Agency	– Relative jurisdictional harmony restored by the Atlantic Accord, Western Accord, Canada–Nova Scotia Off-shore Petroleum Accord and Frontier Energy Policy Resources

223

economic distance in the cheapest possibly way . . .' and that arrangements should be made to market '. . . that portion of Canadian output that cannot be economically used in Canadian refineries in the market that offers the highest return to the producer' (Parliament of Canada, Debates 2928–29). The policy on gas was notably different. No exports were to be permitted until it was demonstrated that '. . . there can be no economic use present or future for that gas in Canada . . .' and, since western gas, delivered by an all-Canadian pipeline was the only reliable source for central Canada '. . . government policy will require that Canadian gas will be used in Canada' (Parliament of Canada 1953).

This statement of policy received national support and provided the framework within which the Transmountain and Interprovincial oil lines and the Westcoast Transmission gas line were approved. It might have been expected that it would also have prepared the way for the relatively rapid approval of the Trans-Canada gas line which, because of its all-Canadian route, some likened to a twentieth-century version of the trans-continental railway because of the manner in which it would link Canada together. Such was not to be the case, however, and the great pipeline debates of 1954–56 were particularly strenuous and acrimonious. The issue was not the concept or the routeing but the financing and the manner in which the government finally dealt with the matter in Parliament.

In 1957 the Liberal government was defeated, in part on the basis of its handling of the Trans-Canada pipeline debates and its willingness to make substantial public funding and guarantees to a company controlled by foreign interests. The new Conservative government appointed a Royal Commission on Energy charged with making recommendations on:

 (i) policies to serve the national interest in relation to the export of energy;
 (ii) policies to regulate the transfer and trade in oil and gas in order to ensure the efficient and economic operation of pipelines in the national interest;
 (iii) the extent of authority, and the administrative structures and procedures of a National Energy Board, and
 (iv) special measures in relation to the Trans-Canada pipeline to safeguard the interests of Canadian producers and consumers of gas (Canada, Royal Commission 1958, pp. 90–1).

The Commission recommended that:

 (i) the export of natural gas '. . . which may from time to time be surplus to the reasonably foreseeable requirements of Canada, be permitted under licence',
 (ii) the tolls and tariffs charged by pipeline companies '. . . should be just and reasonable, non-discriminatory and calculated to yield a fair return on the shareholder's equity . . .',
 (iii) legislation should be enacted to establish a National Energy Board (NEB) which shall report to Cabinet rather than any one Minister of the govern-

ment, be required to approve inter-provincial and foreign trade of all energy, to maintain a data base and, from time to time, prepare comprehensive studies of the energy industry. In its approval procedures, the NEB shall consider the present and future requirements of Canada, the advisability of encouraging downstream development of the energy industry in Canada, the economic feasibility of proposals and the opportunities for Canadians to participate in their financing, engineering, and construction. (Canada, Royal Commission 1958, pp. vii–xiv).

(iv) in order to foster the growth of the Canadian oil industry, exports of oil should be encouraged, and refiners in areas fed by existing pipelines transporting Canadian oil should be required to use that oil. No government action should be taken at this time to ensure the extension of the Canadian oil pipeline system to Montreal. (Canada, Royal Commission 1959, pp. 143–5).

The government acted quickly on the Commission's recommendations by establishing the National Energy Board in 1959, and announcing a National Oil Policy in 1961. This policy was intended '. . . to achieve target levels of production of oil . . .' by means of increased use of Canadian oil in areas west of the Ottawa valley and increased exports to those areas of the United States accessible by existing pipelines (Canada, EMR 1973, pp. 335–6). Increased use west of the Ottawa valley would be achieved by replacing the flow of products from Montreal and the United States into Ontario by products from an expanded Ontario refinery industry using Canadian crude. Increased exports to the United States, permitted as a consequence of Canada being exempted from the import restrictions imposed by the United States government, were to be achieved by more energetic marketing programmes by the industry.

1962-72: Consolidation and regulation The decade 1962-72 started with a Canadian National Oil Policy and an independent regulatory body, the NEB, in place, growing reserves of gas and oil and a world context in which oil was increasingly available and moving in international trade in ever-growing amounts. The Canadian government (which changed back to Liberal in 1963) sought to continue the momentum of the hydrocarbon industry by encouraging exploration which, in turn, meant expanding markets and thus exports. The National Oil Policy permitted the export of oil without licence and the additional Superior–Chicago–Sarnia pipeline, completed in 1968, occasioned little political interest. In fact, oil matters received little attention until the end of the period when concerns over future supplies started to surface.

Focusing on natural gas, the National Energy Board had to deal with three major export issues: the surplus gas available, the price to be charged, and the acceptability of the proposed pipeline facilities (McDougall 1982). Determination of whether surplus gas was available for export involved estimating two major variables, each characterized by a high degree of uncertainty: current and future reserves and future Canadian requirements.

Data on current reserves originated from the producing companies and their exploration divisions had the greatest knowledge of undiscovered resources. Consequently, the NEB was forced to rely on data supplied by the applicants for export licences along with the governments of producing provinces which argued that unless gas could be sold the exploration effort (and hence the discovery rate) would decline. Forecasting Canadian requirements, themselves a function of availability of gas and its price relative to other fuels, could only be undertaken with a relatively high degree of uncertainty which increased as the time horizon of the forecast extended (Chap. 6). Consumer provinces and groups argued for high requirements over a relatively long time (25 years) and that in determining whether a surplus existed, the requirements should be matched against established, not yet-to-be discovered, reserves. In fact, NEB decisions generally favoured export applications and thus the marketing of established reserves, leaving longer-term Canadian needs to be met by as yet undiscovered resources.

During this period the NEB approved Westcoast Transmission's application for price changes and increased volume (but only after protracted deliberations and re-submissions in part accounted for by the United States Federal Power Commission's ruling on price structure), the Great Lakes Transmission application to move additional Canadian gas to south-west Ontario via the United States with an export component and, in 1970, Alberta and Southern's application for the largest single gas export contract in Canada's history to date. McDougall identifies four over-riding principles guiding the NEB to these decisions:

 (i) the maintenance of trust and goodwill between Canada and the United States;
 (ii) gas exports to the United States permit economies of scale in the operation of pipelines and thus contribute to least-cost service to Canadians;
(iii) exports encourage exploration and hence development of reserves as well as contributing directly to regional economic development, and
(iv) the existence and terms of past service should be considered in determining additional supplies to the same area (McDougall 1982, p. 123).

The earlier objective of de-politicizing the energy scene by means of 'expert' NEB rulings generally was achieved during this period. The exception was the decision to permit the Great Lakes Transmission application to route gas destined for Canadian use through the United States. In both the Cabinet and Parliament, considerable concern was again expressed about the potential dangers of United States interventions affecting Canadian supply. These concerns were eventually met by requiring that at least 50 per cent of the total supply of gas to central Canada should move through the all-Canadian Trans-Canada Pipeline.

While these gas related proposals almost completely occupied the attention of the NEB, two other issues were emerging which were to have implications for the 1973–78 period. As early as 1959, exploratory wild-cat drilling had commenced in the Canadian Arctic and, as exploration increased, the federal government sought to establish guidelines to direct the developments. The particular importance of the northern frontier area to the federal government arises from the jurisdiction it has over the Yukon and Northwest Territories and also because of sovereignty issues in relation to the United States hydro-carbon developments in the north slope of Alaska, native land claims, and growing environmental concerns.

In 1961, the Canadian Oil and Gas Lands Regulations were promulgated in an attempt to limit the role of foreign-controlled companies in the Canadian North and, starting in the mid-1960s, steps were taken to further strengthen the federal government's role. In 1966, the federal government purchased a controlling interest in Panarctic Oils Ltd., a consortium of private companies engaged in hydrocarbon exploration in the Arctic Islands. In 1968, an inter-departmental Task Force on Northern Oil Development was formed and, in 1970, soon after the large discovery in Prudhoe Bay, Alaska, preliminary guidelines for the construction and operation of northern pipelines were announced, followed, in 1972, by the *Expanded Guidelines for Northern Pipelines* (Canada, EMR 1973).

The other vitally important issue that emerged in the late 1960s centred on Canada's conventional light oil reserves. The National Oil Policy of 1961 encouraged the unrestricted growth of oil exports. This policy was based on the optimistic views of both established reserves and yet-to-be discovered resources provided by the industry. However, by 1966 the oil discovery rate had started to decline and by 1970 established reserves declined for the first time (Fig. 8.6). This, coupled with the global events of 1973, led to major changes in the context within which Canadian oil and gas policy evolved.

1973-78: Politicization and intervention The year 1973 was the start of an unprecedented decade during which the global energy scene changed rapidly and unpredictably. In Canada, as in most other countries, energy became the foremost national issue, policy formulation became highly politicized and government intervention was greatly increased. The OPEC-determined increase in the price of oil and the selective interruption of supply had four direct effects upon Canada: greater returns to producers, increased cost of production and living, imported oil more expensive than Canadian oil, and public concern over the spectre of supply interruption. In the United States, increasingly large-scale imports of oil and gas (from overseas and Canada) made the American economy even more sensitive to these global changes than the Canadian. The marketing of Alaskan north-slope oil in the remainder of the United States became a high priority leading to the consideration of routes over Canadian territory and through Canadian waters.

Two internal developments were particularly important for the evolution of Canadian policy during this short period. First, it was confirmed that conven-tional light oil reserves had indeed peaked and were insufficient to sustain the contracted level of exports (Canada, NEB 1974). Second, and offering the

227

prospect of a solution to this problem, the northern frontier regions offered tantalizing possibilities of an oil discovery on the scale of the Alaskan north slope (12–15 billion barrels). In addition, the continued dominance of the Canadian hydrocarbon industry by foreign corporations became even more starkly perceived as public attention was directed to the distribution of 'windfall profits', the routeing of crude oil supplies to Montreal refineries and the dominance of corporations as suppliers of data and expertise.

Between 1973 and 1978 three major studies were published by, or under the auspices of, the Federal Department of Energy, Mines and Resources. The first of these was intended:

> . . . to define more clearly the national framework into which provincial studies fit, to identify policy choices which must be made within the federal jurisdiction, and to provide a basis for choice by the government and people of Canada (Canada, EMR 1973, p. iii).

This comprehensive analysis of the national energy scene addressed many issues and suggested numerous policy directions. One of the most significant sections dealt with the concept of economic rent and its division between industry and the federal and provincial governments. The study concluded that the Canadian fiscal regime may no longer result in an appropriate division and should be replaced by one that clearly reflects Canadian conditions rather than those in other parts of the world.

The second study outlined a 'National Energy Strategy for Self-Reliance' and set a number of specific targets aimed at increasing self-reliance (Canada, EMR 1976). The third study assessed Canadian energy prospects beyond 1990 and identified policies and programmes which might be undertaken to adjust to the developments which the authors expected to occur (Gander and Belaire 1978).

The swiftly changing context, however, required rapid reaction and, in 1973, the government took a number of initiatives. The National Oil Program was abolished, oil exports were restricted and made subject to NEB licensing and an oil export charge was imposed. The funds raised by this tax were used to fund an oil import compensation programme so that Montreal refiners would have access to oil at the same basic price as the rest of Canada. In addition, the intention to form a national oil company was announced as well as the extension of the inter-provincial pipeline to Montreal. In 1975, PetroCanada Ltd. was incorporated and the Petroleum Administration Act established federal control over oil and gas pricing. There was also rapid expansion of interest in oil sands, heavy oil and enhanced oil recovery programmes as well as intensification and increased regulation of frontier exploration.

Superimposed on this flood of interventionary moves by the federal government aimed primarily at securing Canada's oil supply and controlling its price (as well as making a start on fundamental changes in the fiscal and organizational structure of the hydrocarbon industry) were the issues arising from northern developments and the search by the United States for ways of marketing Prudhoe Bay oil and gas. Exploration in northern Canada had revealed substantial gas reserves in both the Mackenzie Delta and Arctic Islands and the question was by what route to bring them to the southern markets. When production should start was also of major concern. Alaskan north-slope oil was to be taken westwards to the coast by the Trans-Alaskan Pipeline and thence by tankers to United States ports. Two questions remained: by what route should the large volume of associated gas that would be produced along with the oil be connected to the main United States market, and how should the United States midwest, which had come to rely upon Canadian oil, be supplied in the face of declining Canadian supplies? It was feasible that the marketing of Prudhoe Bay and Mackenzie gas could be linked thus adding complex corporate and operational dimensions to the existing concerns about sovereignty, environmental impact and native land claims. The government turned to boards of inquiry to advise them on these matters. On the gas transportation issue, the Mackenzie Valley Pipeline Inquiry reported nega- tively on the proposed Mackenzie line (Berger 1977) while the Alaska Highway Pipeline Inquiry reported positively on that more southerly route only three months later (Lysyk *et al*. 1977). A third inquiry, the West Coast Oil Ports Inquiry, was appointed to make recommendations about the relative desirabil- ity of proposals to bring oil destined for the United States midwest through British Columbia. This inquiry presented an interim report only as its activities were foreclosed by the withdrawal of the proposals which had led to its establishment (Thompson 1978).

In sum, the period 1973–78 was one of frantic activity during which the Canadian government established control over the pricing and distribution of oil, published three major documents suggesting in increasingly specific terms the way future policy would develop and commissioned three inquiries into the transport of gas and oil from which, however, no action resulted.

1979–83: Comprehensive intervention and The National Energy Program The end of 1978 saw the suspension of oil production in Iran which set in motion a laddered increase of global oil prices from $13/barrel in January 1979 to $34/barrel by mid-1980. At the Economic Summit meeting in June 1980, the leaders of the OECD countries expressed their deep concerns about the price and supply of energy, the implications for inflation, and the level of economic development in their own countries and those of the Third World. Collectively they committed themselves to national programmes which would break the link between economic growth and energy consumption by means of energy conservation, substitution of other energy sources for oil and the mobilization of conventional or unconventional oil supplies outside OPEC.

Inside Canada, after more than a decade of Liberal government, a minority Conservative government was elected in early 1979 only to be defeated (in part because of its ambivalent position on energy policy) and replaced by the Liberals again in early 1980. This was the government that had produced three increasingly specific and widely disseminated studies on energy policy in the previous decade. It was thus poised to react to the dramatic global developments and in October 1980 introduced a comprehensive and strongly interventionist policy package, the implications of which were possibly more revolutionary and far reaching than any other single resource policy initiative in Canadian history (Foster 1982; Doern and Toner 1985). Four major perceived issues were identified in the document setting out the National Energy Program (NEP): (1) the availability and price of crude oil in the international marketplace, (2) the distribution of benefits and burdens within Canada, (3) increased Canadian ownership in the energy industry and (4) the fostering of security, opportunity and fairness (Canada, EMR 1980).

The components of the NEP were numerous and detailed and beyond the scope of this chapter to deal with beyond the items noted in Figure 8.4. The complex of interlocking initiatives was designed to address the four issues listed above by regulation, the creation of an elaborate fiscal structure of taxes and incentives, the introduction of programmes to encourage alternative supplies and bring about improvements in the efficiency of energy use, and to increase the Canadian-controlled share of the hydrocarbon industry. The fiscal initiatives were designed to increase the federal revenues, to encourage frontier exploration and control the price of oil and gas. Programmes to develop alternative sources included joint investment in oil sands projects and to encourage the substitution of gas for oil.

The reaction to the NEP was swift and predictable. The producing provinces were outraged at what they regarded as blatant infringements on their jurisdiction and control of the income from their resource base. The majority of the large companies deplored the 'interventionist strait-jacket' which determined the conditions within which they were to operate, the 'inequity' of the tax structure and the 'confiscatory' nature of some of the programmes aimed at Canadianization of the industry. Consumer reaction was generally more positive. They perceived themselves to be protected from the dramatic global price increases and, to many, the strengthening of the role of Petro-Canada Ltd was appealing (Doern and Toner 1985; Carmichael 1983; Watkins and Stabback 1981). Some of the objections were sufficiently strong and substantive to require government reaction and, in 1982, a revised document was published which was both a review of the impact of the NEP and a revision of some of its elements (Canada, EMR 1982).

The NEP reflected the philosophy of the Liberal government of the day and that of its major architects, a group of senior civil servants some of whom had been specifically recruited to devise the programme (Foster 1982). It was perceived by these policy makers as providing Canadians with an energy future which was secure, fair and made-in-Canada. However, changes in both the global and domestic context were to make realization of these expectations difficult. Globally, demand for energy started to decrease (as the result of the combined effects of recession, effective conservation programmes and high prices), the price of oil stabilized and then started to decrease and the power of OPEC started to wane. In the United States, deregulation of the oil and, more slowly, the gas industry had significant marketing implications for Canadian gas and oil. Domestically, the big oil discovery in the frontier areas continued to elude the costly exploration efforts and public opinion was swinging away from controls and interventionism and towards entrepreneurial freedom and the competitive environment created by market forces.

1984–87: Disengagement and deregulation As the mid-1980s approached these trends accelerated. Oil prices continued to decline and tensions increased among members of OPEC, and the United States market for gas was increasingly difficult to penetrate both as a result of deregulation and the emergence of a general protectionist attitude. In Canada, shut-in gas reserves increased, relations between the two levels of government further deteriorated and the mood of the populace moved towards a desire for political change. In September, 1984 a Progressive Conservative government was elected which was committed to:

(i) development of energy resources as an engine of growth to the benefit of all of Canada;
(ii) self-sufficiency and energy security;
(iii) enhanced Canadian participation;
(iv) fair treatment for consumers and producers, and
(v) cooperation between federal and provincial governments and industry to produce a stable planning environment. (Canada, Office of the Leader of the Opposition 1984.)

On the face of it, these goals seem very similar to those of the previous government. However, there were two major differences in the manner in which policy formulation was to be approached both reflecting the disposition of the Progressive Conservative Party and a shift in public attitudes. The first was the adoption of a consultative process between the two levels of government, industry and public interest groups and, second, the dismantling of the web of regulatory controls of the NEP and a return to the operation of the market as the driving force behind energy sector decisions. During 1985 most of the legislation and implementation programmes of the NEP were scrapped and replaced by a series of agreements on jurisdictional and deregulation issues (Canada, EMR 1985a, b, c and d). The important changes arising from these agreements were the establishment of a considerable measure of accord between governments and industry, the deregulation of oil and gas prices, a new fiscal regime including major revision to the tax system and NEB regulations concerning the pricing of exports, a new direction for the development of energy in the frontier regions and financial support for heavy oil upgrading facilities in western Canada.

The NEP of 1980 may be seen as an attempt by a federal government to take control of the major forces shaping the Canadian energy system and to regulate them in order to develop a system with pre-conceived characteristics. The post-1984 Progressive Conservative government, while maintaining some relatively clear objectives with respect to the Canadian energy system, explicitly chose to remove the control and regulation of the major variables and let entreprenurial decisions taken in the context of market forces shape the energy industry. As proponents of the NEP discovered, even a closely regulated system was susceptible to disruption by forces which were not controlled such as, internally, public opinion and, externally, global oil prices. The *laissez-faire* strategy of the mid-1980s by definition permitted internal and external market forces to operate. Policy interventions were to be kept to a minimum and, once introduced, used only to steer market forces in desired directions.

The most important external development during this period was the collapse of world oil prices to less than $12/barrel and, by mid-1987, their recovery to the $18–20 range. For much of the time these changes resulted in Canadian oil being more expensive than off-shore oil in central and eastern Canadian markets. Furthermore, they decreased the incentive for oil exploration and new investment in production facilities that would yield oil which could be marketed at only marginal or negative returns. Exports of oil did not face the marketing problems constraining gas deliveries. Although volumes were less than earlier years, exports increased somewhat after 1983 in part due to the absence of refineries in western Canada able to process the heavy oil which constituted an increasing proportion of Canadian production.

As a consequence of these developments, the upstream end of the Canadian oil and gas industry became depressed which, coupled with operating in a new, deregulated context placed considerable strain on the industry. After a short period of almost euphoric reaction to the demise of the NEP, the industry and, to some extent, the producing provinces, appealed for assistance from the federal government. For the most part the appeals fell short of asking for a return of controls but focused on alterations to the fiscal structure worked out in cooperation with the producers.

After three decades of large scale investment in new production, transportation and processing facilities in which the spatial structure of the hydrocarbon industry was established, relatively little facility investment occurred during this period. Despite the furore of the previous decade, no delivery systems were built from the frontier areas but construction on the Quebec and Maritime extension of the Trans-Canada pipeline system continued and the capacity of the Interprovincial oil line was increased in part to accommodate the increasing proportion of slow-moving heavy oil.

Of more concern than new investment was the development of new marketing strategies for gas in response to deregulation, streamlining approval procedures for gas exports and modifying the conditions under which oil pipelines were licensed so as to permit them to transport petroleum products and natural gas liquids as well as crude oil.

Geography and public policy

The relations between geography and public policy are complex and interactive. It is helpful for discussion purposes to adopt some working definitions of the two terms and then express the interactive nature of their connection in a simplified way as two uni-directional relations.

Working definitions

Geographers, political scientists and policy analysts will all have accepted definitions of geography and public policy which, very likely, will differ from one another. To avoid the kind of unproductive discussion that can arise when the meaning of terms is left unspecified let 'geography' in this context be taken to refer to a number of variables clustered around two concepts, spatial structure and the bio-physical environment. The elements of spatial structure consist of the familiar spatial location, distribution, association and interaction of phenomena, and the elements of biophysical environment are the terrestrial, hydrologic, atmospheric and biotic components of an area which interact in an holistic and dynamic relationship.

The term 'public policy' has been developed in Figure 7.1. The framework presented there portrays the four major components of public policy as perceived issues, formulation, elements and results which evolve within a context. The *context* consists of the state of the system under consideration and its controlling variables and the prevailing ideologies, both political and societal.

Depending upon the ideology of the group concerned, some aspects of the state of the system will be *perceived as issues* leading to the expectation that some conscious decisions should be taken to change the state of the system (i.e. manage it). *Formulation* of the policy required to accomplish the change involves a number of participants being brought together in some process aimed at producing a set of policy *elements*. These elements include, as a minimum, statements on objectives, strategies, programmes and administration. Whether the desired *results* have been achieved can only be determined by monitoring and evaluation of the information revealed. On the basis of the evaluation, adjustments (feedback) may be made either to fine-tune the prevailing policy or to shift direction more radically.

Inter-relations

The inter-relations between geography and public policy may initially and simplistically be expressed by assuming first one then the other to be the dependent variable. Thus one may explore the influence of geographic variables upon public policy and vice versa.

Geographic influences on Canadian public policy   The basic spatial characteristic with which all Canadian energy policy makers have had to contend is the discordance between the spatial distribution of domestic fuel resources and major domestic market areas. In addition to this fundamental pattern are two others of underlying importance to Canadian fuel policy: the spatial proximity

233

of some United States fuel supplies to Canadian markets (persistently for coal and formerly for oil and gas) and the proximity of United States markets for oil and gas to Canadian supplies. These spatial imperatives largely account for the persistence of transportation (routeing, costs) and trade (decision to export, amount, conditions) matters as perceived issues for Canadian policy makers to deal with.

Another consideration of more recent significance is the spatial intervention of Canadian territory and waters between Alaska and the remainder of the United States. American interests, both governmental and corporate, in transporting oil and, particularly, gas from the Alaskan north slope through Canada led Canadian policy makers in the 1970s into a preoccupation with the routeing and regulation of proposals, none of which has yet (1987) been developed. The presence of these issues on the policy agenda was also related to the marketing of gas from the northern Canadian frontier.

Before leaving the spatial aspect of geographic influences on Canadian energy policy it is important to recognize the influence of one other persistent distributional characteristic, the distribution of population and, thus, of political power. Central Canadian (Ontario and western Quebec) voters, representing the fuel deficit/major market area (the 'acute fuel area' in the debates on coal policy), constitute approximately 60 per cent of the national electorate and, thus, tend to have the dominant political voice in the country. National policy makers have had to recognize this fact since Confederation.

Environmental considerations shape public policy in two distinct but related ways. First, the environment influences the difficulty and cost of constructing and operating the components of energy systems and thus the viability of one proposal over another. For example, the all-Canadian route for western gas to central Canada must pass through the Canadian Shield. Costs per kilometre of pipeline construction are very high through this hard rock area, posing major financial problems for the proponents of the route which were overcome by the contentious policy decision to provide public capital and loan guarantees. Similarly, the harsh and hazardous environmental characteristics of the frontier areas result in very high exploration costs which required special measures such as the Petroleum Incentive Program to achieve the objectives of the NEP.

The impact of energy systems upon the environment has become a major perceived issue in energy policy. The maintenance of environmental quality and the preservation of environmental complexes and components of environments which are considered to be unique, representative and of heritage value have become increasingly important to the Canadian public. The national government has had to take more and more account of these values in the regulations and procedures governing the operation of energy facilities, especially in the western Arctic (e.g. Guidelines for Northern Pipelines). Indeed, the decision not to permit the proposed Mackenzie Valley Pipeline was seen by many as a clear illustration of the role of environmental considerations as a major variable in shaping public policy.

Public policy influences on geography Exploration of the influence of public policy upon the spatial structure of the energy industry and the character of the environment in Canada reveals some clear and direct effects and others that are more subtle and indirect. Policy decisions leading to the provision of incentives (or disincentives) for exploration in selected areas can affect the future spatial distribution of reserves. More directly, the spatial distribution of fossil fuel production can be significantly changed if conditional reserves are converted to current status (and vice versa) by policy decisions. Capital grants to oil sands projects, financial assistance to pipelines and the award of export licences are examples of public initiatives which had clearly identifiable consequences for the spatial structure of energy supply in Canada.

Perhaps less obvious but of considerable importance to regional economic development is the influence of public policy on the location of the downstream components of energy systems (e.g., the petroleum refining and petro-chemical industries) and industries which are large energy consumers for process purposes (e.g. the integrated iron and steel industry). For example, the National Oil Policy of 1961 led to the expansion of the oil refining industry in Ontario and the later licensing of pipelines to carry natural gas liquids did much to sustain the petro-chemical industry of south-west Ontario just as the application before the NEB in 1987 to permit the IPL to carry NGLs to Montreal might revive that city's petro-chemical industry.

It is difficult to point to specific examples of the influence of federal fossil fuel policy upon the quality and preservation of environment. Federal jurisdiction is paramount only in the territories in which there has been considerable exploration activity, mostly conducted with a great deal of attention to avoiding severe environmental disruption. Within the individual provinces federal jurisdiction is restricted by the terms of the Constitution so that a search for examples of the influence of federal energy policy on environmental quality is not likely to be fruitful. Furthermore, it is environmental policy rather than energy policy that is likely to be more significant in this respect.

The descriptive overview of the evolution of the Canadian fossil fuel industry and the associated federal government policy presented in this chapter serves to illustrate the interaction between geography and public policy. The geographical characteristics of discordance between supply and consuming regions, the accessibility of markets in the USA to Canadian supplies (and vice versa) and the patterns of US–Canadian territorial distribution in relation to potential energy transportation routes constitute a set of pervasive spatial relations with which Canadian policy makers have had to contend for over a century. When approached from such often conflicting objectives as national unity, energy independence, equitable distribution of costs and benefits and sovereignty, successive Canadian governments have dealt with these spatial imperatives by policies which reflect the evolving political and social realities of the Canadian scene as well as changing events in the USA and beyond.

APPENDIX A

Units and conversions

I. Energy and power

A. Units

British	American	International
British thermal unit (BTU)		kilocalorie or Calorie
foot-pound		joule
horsepower		watt

B. Energy conversions

one		equivalent to
kilocalorie	3.97	BTU
kilocalorie	3.08×10^3	foot-pounds
kilocalorie	4184	joules
kilocalorie	1.16×10^{-3}	kilowatt-hour
kilocalorie	1.55×10^{-3}	horsepower-hour
BTU	0.252	kilocalorie
BTU	776	foot-pounds
BTU	1054	joules
BTU	2.93×10^{-4}	kilowatt-hour
BTU	3.93×10^{-4}	horsepower-hour
kilowatt-hour	860	kilocalories
kilowatt-hour	3413	BTU
kilowatt-hour	3.61×10^6	joules
kilowatt-hour	2.65×10^6	foot-pounds
kilowatt-hour	1.34	horsepower-hours
joule	2.39×10^{-4}	kilocalorie
joule	9.48×10^{-4}	BTU
joule	0.735	foot-pound
joule	2.78×10^{-7}	kilowatt-hour
joule	3.72×10^{-7}	horsepower-hour

C. Power conversions

one		equivalent to
kilowatt	1000	watts (joules per second)
kilowatt	736	foot-pounds per second
kilowatt	0.239	kilocalorie per second
kilowatt	0.948	BTU per second
kilowatt	1.34	horsepower
horsepower	550	foot-pounds per second
horsepower	0.746	kilowatt
horsepower	0.179	kilocalorie per second
horsepower	0.706	BTU per second

From: Fowler (1984) 632–3

II. Commodity measures

A. Units

	British	American	International
Mass	pounds (lb)		kilograms (kgm)
	long ton	short ton	metric ton (tonne)
Volume	cubic feet		cubic metres
	gallons (Imperial)	gallons (U.S.)	litres
		barrel	

B. Conversions

Mass

	Metric tonnes	Long tons	Short tons	Pounds	Kilograms
Metric tonne*	—	0.984	1.102	2205	1000
Long ton	1.016	—	1.120	2240	1016
Short ton	0.907	0.893	—	2000	907.2
Pound	0.000453	0.000446	0.0005	—	0.4536

* In converting crude oil to kilolitres or cubic metres, one metric tonne is approximately equal to 1.16 kilolitres, or 1.16 cubic metres

Appendix A

Volume

	Barrels	Imperial gallons	US gallons	Kilolitres
Barrel	—	34.973	42	0.159
Imperial gallon	0.0286	—	1.201	0.00455
US gallon	0.0238	0.833	—	0.00379
Kilolitre*	6.2898	219.97	264.17	—

* In converting crude oil to tonnes one kilolitre is approximately equal to 0.863 tonnes

	Cubic feet	Cubic metres
Cubic foot	—	0.0283
Cubic metre	35.315	—

From: Crabbe and McBride 1978, *The World Energy Book*

III. Energy content

Approximate calorific values for different energy sources

	Therms per tonne	GJ/tonne		Therms per tonne	GJ/tonne
Crude oil (average)	425	45	Anthracite	315	33
Petroleum gases	500	53	Bituminous coal (average)	275	29
Liquefied petroleum gas	470	50	Sub-bituminous coal (average)	235	25
Naphtha	455	48	Brown coal and lignite	140	15
Motor spirit	445	47			
Burning oil	440	46	Peat	75	8
Gas/Diesel oil	430	45	Coke	265	28
Fuel oil	405	43			

1 GJ is approximately equal to 9.5 therms

Calorific equivalents – conversion factors (multiply by the figure shown)

To \ From	Million tonnes oil equivalent	Million tonnes coal equivalent	Million therms	Terawatt hours (thermal)	Terawatt hours (electrical)	Terajoules
Million tonnes oil equivalent (mtoe)	1	0.60	0.0024	0.0800*	0.2667**	2×10^{-5}
Million tonnes coal equivalent (mtce)	1.67	1	0.0039	0.1335*	0.4450**	4×10^{-5}
Million therms (m th)	425	255	1	34.13	114	0.0095
Terawatt hours (thermal) (TWh[t])	12.50	7.49	0.0293	1	3.33	0.0003
Terawatt hours (electrical) (TWh[e])	3.75	2.25	0.0088	0.30	1	0.0001
Terajoules (TJ)	44,800	26,900	105.5	3600	12000	1

* The quantity of fuel equivalent to 1 TWh of energy
** The quantity of fuel required to generate 1 TWh of electricity

Approximate calorific equivalents for conversion into million tonnes of oil

One million tonnes of oil equals approximately	Heat units and other fuels expressed in terms of million tonnes of oil	
Heat Units		million tonnes of oil
43 million million BTU	10 million million BTU approximates to	0.24
425 million therms	100 million therms approximates to	0.24
44800 Terajoules	10,000 Terajoules approximates to	0.22
Solid Fuels		
1.67 million tonnes of coal	1 million tonnes of coal approximates to	0.60
3.04 million tonnes of lignite	1 million tonnes of lignite approximates to	0.33
5.67 million tonnes of peat	1 million tonnes of peat approximates to	0.18
Natural Gas (1 ft^3 = 1000 BTU) (1 m^3 = 37.1 MJ)		
1.17 thousand million m^3	1 thousand million m^3 approximates to	0.85
41.4 thousand million ft^3	10 thousand million ft^3 approximates to	0.24
113 million ft^3/day for a year	100 million ft^3/day for a year approximates to	0.88

From: Crabbe and McBride 1978, *The World Energy Book*

Appendix A

IV. Large numbers

Word	Prefix	Exponential notation	Symbol
Ten	deka	10^1	da
Hundred	hecto	10^2	h
Thousand	kilo	10^3	K
Million	mega	10^6	M*
Billion	giga	10^9	G
Trillion	tera	10^{12}	T
Quadrillion	peta	10^{15}	P**
	exa	10^{18}	E

* M is sometimes used to denote mille (thousand)
** Q for quad is sometimes used instead of P

REFERENCES

Aharari, M. E. (1986) *OPEC, the Falling Giant*, University Press of Kentucky, Lexington, Kentucky.

Al-Chalabi, F. J. (1980) *OPEC and the International Oil Industry: A Changing Structure*, Oxford University Press, Oxford.

Ali, S. R. (1986) *Oil, Turmoil and Islam in the Middle East*, Praeger Publishers, New York.

American Petroleum Institute (1986) *Basic Petroleum Data Book: Petroleum Industry Statistics*, vol. **VI** (12): API, Washington DC.

Amman, F. and R. Wilson (eds) (1980) *Energy Demand and Efficient Use*, Plenum Press, New York.

Australia, Department of Resources and Energy (1984) *Energy Demand and Supply, Australia, 1960–61 to 1982–83*, Australian Government Publishing Services, Canberra.

Ayoub, A. (ed.) (1979) *Energy: International Cooperation or Crisis?*, University of Laval Press, Quebec.

Barker, B. (ed.) (1981) *Earth's Renewable Resources: A Special Report on Technology Development*, *EPRI Journal* **6** (10), whole issue.

Beca, Carter, Hollings and Ferner in association with R. A. Shaw (1979) *Greater Auckland Commercial Sector Energy Analysis,* Report No. 45, New Zealand Energy Research and Development Committee, Auckland.

Beaujean, J. M. and J. P. Charpentier (ed.) (1978) *A Review of Energy Models*, RR 78–12, International Institute of Advanced Systems Analysis, Laxenburg, Austria.

Berger, T. R. (1977) *Northern Frontier, Northern Homeland*, Report of the Mackenzie Valley Pipeline Inquiry, Volumes I and II, Supply and Services Canada, Ottawa.

Bernstein, H. M. (1975) *A Study of the Physical Characteristics, Energy Consumption and Related Institutional Factors in the Commercial Sector*, National Technical Information Service, Springfield, Virginia.

Blunden, J. (1985) *Mineral Resources and Their Management*, Longman, London.

Bohi, D.R. and M. A. Toman (1983) 'Understanding nonrenewable resource supply behaviour', *Science*, 219 No. 4587, 927–30.

Bohi, D.R. and M. B. Zimmerman (1984) 'An update on econometric studies of energy demand behavior', *Annual Review of Energy* **9**: 105–54.

Bradshaw, M. J. (1987) *Soviet–Pacific Basin Trade: A Canadian Perspective*, Working Paper No. 29, Institute of Asian Research, University of British Columbia, Vancouver BC.

241

References

British Petroleum (1979, 1986 and 1987) *BP Statistical Review of World Energy*, British Petroleum, London.

Broadman, H. G. (1985) 'Incentives and constraints on exploratory drilling for petroleum in developing countries', *Annual Review of Energy* **10**: 217–49.

Budnitz, R. J. and J. P. Holden (1976) 'Social and environmental costs of energy systems,' *Annual Review of Energy* **1**: 553–80.

Bush, R. P. and A. T. Chadwick (1979) *A Disaggregated Model of Energy Consumption in UK*, Report No. 8, ETSU, Harwell, Oxon.

Cain, B. and P. Nevin (1982) *International Energy Comparisons: A View of Eight Industrialized Countries*, Discussion Paper No. 222, Economic Council of Canada, Ottawa.

Calzonetti, F. J. and B. D. Solomon (eds) (1985) *Geographical Dimensions of Energy*, D. Reidel Publishing Company, Dordrecht.

Canada, Department of Energy, Mines and Resources (EMR) (1973) *An Energy Policy for Canada: Phase I*, Volumes I and II, Information Canada, Ottawa.

Canada, Department of Energy, Mines and Resources (1976), *An Energy Strategy for Canada: Policies for Self-Reliance*, Information Canada, Ottawa.

Canada, Department of Energy, Mines and Resources (1977) *Energy Demand Projections: A Total Energy Approach*, Report ER 77–4, Supply and Services, Ottawa.

Canada, Department of Energy, Mines and Resources (1978) *Oil and Natural Gas Industries in Canada, 1978*, Report ER 78–2, Supply and Services, Ottawa.

Canada, Department of Energy, Mines and Resources (1980) *The National Energy Program: 1980*, Supply and Services, Ottawa.

Canada, Department of Energy, Mines and Resources (1981) *An Introduction to the Energy Cascading Potential in Canadian Industry*, Conservation and Renewable Energy Branch, EMR, Ottawa.

Canada, Department of Energy, Mines and Resources (1982) *The National Energy Program: Update 1982*, Supply and Services, Ottawa.

Canada, Department of Energy, Mines and Resources (1985a) *The Atlantic Accord*, Office of the Minister, EMR, Ottawa.

Canada, Department of Energy, Mines and Resources (1985b) *The Western Accord*, Office of the Minister, EMR, Ottawa.

Canada, Department of Energy, Mines and Resources (1985c) *Canada's Energy Frontiers: A Framework for Investment and Jobs*, Supply and Services, Ottawa.

Canada, Department of Energy, Mines and Resources (1985d) *Agreement on Natural Gas Markets and Prices*, Office of the Minister, EMR, Ottawa.

Canada, Department of Energy, Mines and Resources (1986) *Canada–Nova Scotia Offshore Petroleum Resources Accord*, Supply and Services, Ottawa.

Canada, Department of Energy, Mines and Resources (1987) *Energy in Canada: An Overview*, Energy, Mines and Resources, Ottawa.

Canada, Federal–Provincial Task Force (1986) *Western Canadian Low-Sulphur Coal: Its expanded Use in Ontario*, Supplies and Services, Ottawa.

Canada, House of Commons (1921) *Future Fuel Supply of Canada*, Official Report of Evidence to Special Committee of the House of Commons, King's Printer, Ottawa.

Canada, House of Commons (1923) *Canadian Fuel Supply*, Minutes of Select Standing Committee on Mines and Minerals, King's Printer, Ottawa.

Canada, Ministry of State for Urban Affairs (1977) *Habitat and Energy in Canada*, Canada Mortgage and Housing Corporation, Ottawa.

Canada, National Energy Board (NEB) (1968) *Energy Supply and Demand Balances, 1955–1967*, Queen's Printer, Ottawa.

Canada, National Energy Board (1969) *Energy Supply and Demand in Canada and Export Demand for Canadian Energy, 1966–1980*, Queen's Printer, Ottawa.

Canada, National Energy Board (1974) *Report to the Honourable Minister of Energy,*

Mines and Resources in the Matter of the Exportation of Oil, National Energy Board, Ottawa.

Canada, National Energy Board (1986) *Annual Report, 1986*, National Energy Board, Ottawa.

Canada, Office of the Leader of the Opposition (1984) *Progressive Conservative Agenda for Government Policy Area: Energy*, House of Commons, Ottawa.

Canada, Royal Commission (1947) *Report of the Royal Commission on the Coal Industry in Canada*, King's Printer, Ottawa.

Canada, Royal Commission (1958) *First Report of Royal Commission on Energy*, Queen's Printer, Ottawa.

Canada, Royal Commission (1959) *Second Report of Royal Commission on Energy*, Queen's Printer, Ottawa.

Canada, Royal Commission (1960) *Report of the Royal Commission on Coal*, Queen's Printer, Ottawa.

Canadian Imperial Bank of Commerce (Annual 1955–1978) *Canadian Petroleum Highlights*, Petroleum and Gas Department, Calgary.

Carmichael, E. A. (1983) *Lessons from the National Energy Program*, Publication No. 25, C. D. Howe Institute, Montreal.

Carroll, W. F. (1947) *Report of the Royal Commission on Coal, 1946*, King's Printer, Ottawa.

Carter, D. B., T. H. Schmudde and D. M. Sharpe (1972) *The Interface as a Working Environment: A Purpose for Physical Geography*, Technical Paper No. 7, Commission on College Geography, American Association of Geographers, Washington DC.

Carter, L. J. (1978) '*Amoco Cadiz* incident points up the elusive goal of tanker safety', *Science* **200**: 514–16.

Champness, M. V. D. (1981) 'The outlook for transport of oil', paper presented to *World Transport of Energy Conference*, Royal Overseas League, London.

Chapman, D. (1983) *Energy Resources and Energy Corporations*, Cornell University Press, Ithaca, NY.

Chapman, J.D. (1961) 'A geography of energy: an emerging field of study', *The Canadian Geographer*, **V**: 10–15.

Chapman, J. D. (1976) 'Geographers and energy', in B. M. Barr (ed.), *New Themes in Western Canadian Geography: The Langara Papers*, B.C. Geographical Series No. 22, Tantalus Research, Vancouver.

Chapman, K. (1976) *North Sea Oil and Gas, A Geographical Perspective*, David and Charles, Newton Abbot.

Chardonnet, J. (1962) *Les Sources d'Energie*, Editions Sirey, Paris.

Cockshutt, E. P. (1973) 'Energy in transportation', *Quarterly Bulletin*, No. 3, Department of Mechanical Engineering, National Research Council, Ottawa, 23–32.

Committee on Mineral Resources and the Environment (1975) *Mineral Resources and the Environment*, National Academy of Sciences, Washington DC.

Cook, E. (1971) 'The flow of energy in an industrial society', *Scientific American* **224**: 135–44.

Cook, E. (1976) *Man, Energy and Society*, W. H. Freeman, San Francisco.

Cook, E. (1977) *Energy: The Ultimate Resource?*, Resource Paper No. 77–4, Commission on College Geography, Association of American Geographers, Washington DC.

Copp, D. and D. Levy (1982) 'Value neutrality in the techniques of policy analysis: risk and uncertainty', *Journal of Business Administration* **13** (1 and 2): 161–90.

Cowhey, P. F. (1985) *The Problems of Plenty: Energy Policy and International Politics*, University of California Press, Berkeley and Los Angeles.

Crabbe, D. and R. McBride (1978) *The World Energy Book: An A–Z Atlas and Statistical Source Book*, Nichols Publishing Co., New York.

References

Craig, P. P., J. Darmstadter and S. Rattien (1976) 'Social and institutional factors in Energy conservation', *Annual Review of Energy* **I**: 535–51.

Curran, D. W. (1973) *Géographie mondiale de l'énergie*, Masson, Paris.

Darmstadter, J., P. D. Teitelbaum and J. G. Polach (1971) *Energy in the World Economy: A statistical review of trends in output, trade and consumption since 1925*, Johns Hopkins University Press, Baltimore.

Darmstadter, J. (1975) *Conserving Energy: Prospects and Opportunities in the New York Region*, Johns Hopkins University Press, Baltimore.

Davey, W. G. (1987) 'Energy issues and policies in Eastern Europe', *Energy Policy* **15**, 1, 59–72.

Dienes, L. and T. Shabad (1979) *The Soviet Energy System: Resource Use and Politics*, J. Wiley, Washington.

Dienes, L. (1981) 'Energy conservation in the USSR', Chapter V in *Report of US Congress Joint Economic Committee*, Washington DC, 101–19.

Dobson, W. (1981) *Canada's Energy Policy Debate*, Publication No. 23, C.D. Howe Institute, Montreal.

Doern, G. B. and P. Aucoin (eds) (1979) *Public Policy in Canada: Organization, Process and Management*, Gage Publishing, Toronto.

Doern, G. B. and G. Toner (1985) *The Politics of Energy: the Development and Implementation of the NEP*, Methuen Publications, Agincourt, Ontario.

Dominion Bureau of Statistics (DBS) (1957) *Energy Sources in Canada: Commodity Statements for 1926, 1929, 1933 and 1939*, Reference Paper No. 74, Queen's Printer, Ottawa.

Dunkerley, J. (1977) *International Comparison of Energy Consumption*, Resources for the Future, Research Paper R–l0, Johns Hopkins University Press, Baltimore.

Dunkerley, J. (1980) *Trends in Energy Use in Industrial Societies: An Overview*, Resources for the Future, Research Paper R-19, Johns Hopkins University Press, Baltimore.

Dunkerley, J., W. Ramsay, L. Gordon and E. Cecelski (1981) *Energy Strategies for Developing Nations*, Johns Hopkins University Press, Baltimore.

Economic Council of Canada (1985) *Connections: An Energy Strategy for the Future*, Supply and Services, Ottawa.

Eden, R., M. Posner, R. Bending, E. Crouch and J. Stanislaw (1982) *Energy Economics: Growth, Resources and Policies*, Cambridge University Press, Cambridge.

Editors, The (1985) *Energy and Environment: The Unfinished Business*, Congressional Quarterly, Washington, DC.

Edmonds, J. and J. M. Reilly (1985) *Global Energy: Assessing the Future*, Oxford University Press, New York.

Energy Policy (1975) **3**, No. 4, Special issue on Energy Analysis.

Fesharaki, F. and D. T. Isaak (1983) *OPEC, the Gulf and the World Petroleum Market: A Study in Government Policy and Downstream Operations*, Westview Press, Boulder, Colorado.

Foster, P. (1982) *The Sorcerer's Apprentices: Canada's Super-Bureaucrats and the Energy Mess*, Collins, Toronto.

Fowler, J. M. (1975) *Energy and the Environment*, McGraw Hill, New York.

Fowler, J. M. (1984) *Energy and the Environment*, 2nd edition. McGraw Hill, New York.

Fuller, J.F. (1984) 'The increasing leverage of geophysics', *Exploration and Economics of the Petroleum Industry* **22**: 361–403.

Gander, J. E. and F. W. Belaire (1978) *Energy Futures for Canadians: Long Term Energy Assessment Program*, Supply and Services Canada, Hull.

Gates, D.M. (1985) *Energy and Ecology*, Sinauer Associates, Massachusetts.

George, P. (1950) *Géographie de L'Energie*, Libraire de Medicis, Paris.

Gerlach, L. P. (1982) 'Dueling the devil in the energy wars', Chapter 4 in P. P. Craig and M. D. Levine (eds), *Decentralized Energy*, Westview Press, Boulder, Colorado.

Ghosh, A. (1983) *OPEC, the Petroleum Industry and US Energy Policy*, Greenwood Press, Westport, Connecticut.

Gilliland, M. W. (1975) 'Energy analysis and public policy', *Science* 189: 1051–6.

Gold, T. (1985) 'The origin of natural gas and petroleum, and the prognosis for future supplies', *Annual Review of Energy* 10: 53–77.

Gorst, E. (1987) 'Oil export rise in 1986', *Petroleum Economist*, LIV, 3: 93–5.

Greenberger, M. (1977) 'Closing the circuit between modelers and decision makers', *EPRI Journal* 8: 6–13.

Greene, D. L. (1979) 'State differences with demand for gasoline: an econometric analysis', *Energy Systems and Policy* 3: 191–212.

Guiness, P. (1979) 'Nuclear power and the American energy crisis', *Geography* 64: 12–16.

Gustafson, T. (1983) 'Soviet energy policy', in *Soviet Economy in the 1980s: Problems and Prospects*, US Congress, Joint Economic Committee, Washington DC, 431–56.

Gustafson, T. (1985) 'The origins of the Soviet oil crisis, 1970–1985', *Soviet Economy* 1, 2: 103–35.

Guyol, N. B. (1971) *Energy in the Perspective of Geography*, Prentice Hall, Englewood Cliffs, NJ.

Häfele, W. (1981) *Energy in a Finite World: A Global Systems Analysis*, Ballinger, Massachusetts.

Hannon, B. and J. R. Broderick (1982) 'Steel recycling and energy conservation', *Science* 216: 485–91.

Harder, E. L. (1982) *Fundamentals of Energy Production*, Wiley, New York.

Hare, F.K. (1953) *The Restless Atmosphere*, Hutchinson University Library, London.

Hare, F. K. and K. Hewitt (1973) *Man and Environment: Conceptual Frameworks*, Resource Paper No. 20, Commission on College Geography, Association of American Geographers, Washington DC.

Hartshorn, J. E. (1980) 'From multinational to national oil: the structural change', Journal of Energy and Economic Development, V: 207–20.

Hay, J. and K. J. Hanson (1985) 'Evaluating the solar resources: a review of problems resulting from temporal, spatial and angular variations', *Solar Energy* 34 (2): 151–61.

Hirsch, R. L. (1987) 'Impending United States energy crisis', *Science* 235: 1467–73.

Hoffman, T. and Johnson, B. (1981) *The World Energy Triangle: A Strategy for Cooperation*, Ballinger, Cambridge, Mass.

Hoffman, K. C. and D. O. Wood (1976) 'Energy system modelling and forecasting', *Annual Review of Energy* I: 423–53.

Holder, G. D. (1984) 'The potential of natural gas hydrates as an energy resource', *Annual Review of Energy* 9: 427–45.

Holdren, J. P., G. Harris and I. Mintzer (1980) 'Environmental aspects of renewable Energy sources', *Annual Review of Energy* 5: 241–91.

Hough, G. V. (1985) 'Record production and reserves', *Petroleum Economist* LII (8): 271–3.

Hough, G. V. (1987) 'Long-term view of energy policy', *Petroleum Economist* LIV (2): 55–7.

Hubbert, M. K. (1962) *Energy Resources*, Publication 1000-D, National Academy of Sciences, National Research Council, Washington DC.

Huettner, D. A. (1976) 'Net Energy analysis: an economic assessment, *Science* 192: 101–4.

Hughes, B. B. *et al.* (1985) *Energy in the Global Arena: Actors, Values, Policies and Futures*, Duke University Press, Durham NC.

References

International Energy Agency (1979) *Workshop on Energy Data of Developing Countries. Vol. II, Basic Energy Statistics and Energy Balances of Developing Countries, 1967–1977*, OECD, Paris.

International Energy Agency (1982) *World Energy Outlook*, OECD, Paris.

International Energy Agency (1983a) *Energy Balances of OECD Countries, 1971–1981*, OECD, Paris.

International Energy Agency (1983b) *Coal: Environmental Issues and Remedies*, OECD, Paris.

International Energy Agency (1986) *Energy Policies and Programmes of IEA Countries*, OECD, Paris.

International Petroleum Encyclopedia, 1981, Pennwell Publishing, Tulsa.

Isserman, A. M. (1977) 'Some policy implications of spatial variations in fuel consumption by manufacturing activities', *Economic Geography* **53**: 45–54.

Kakela, P. J. (1978) 'Iron Ore: energy, labor and capital changes with technology', *Science* **202**: 1151–7

Kalma, J. D. (1976) *Sectoral Use of Energy in Australia*, Technical Memorandum 76/4, CSIRO, Canberra.

Kalma, J. D. and K. J. Newcombe (1976) 'Energy use in two large cities: a comparison of Hong Kong and Sydney, Australia', *International Journal of Environmental Studies* **9**: 53–64.

Kalma, D., M. Johnson and K. J. Newcombe (1978) 'Energy use and the atmospheric environment: Part I. Inventory of air pollution emissions and prediction of ground level concentrations of SO^2 and CO', *Urban Ecology* **3**: 29–57.

Kaser, M. (1986) 'The energy crisis and Soviet economic prospects 1986–1990', *The World Today,* June, 1986, 91–2.

Katz, J. E. (1984) *Congress and National Energy Policy*, Transaction Books, New Brunswick, New Jersey.

Keyfitz, N. *et al.* (1983) *Global Population, 1975–2075*, Research Memorandum ORAU/IEA–83-6(M), Institute for Energy Analysis, Oak Ridge, Tennessee.

Kroncher, A. (1985) 'The future of Siberian gas in danger', *Radio Liberty Research Bulletin* RL 31/85, January 30, 1–3.

Levine, D. (1977) *Project Interdependence: US and World Energy Outlook through 1990*, Congressional Research Service, Library of Congress, Washington DC.

Lihach, N. (1982) 'New connections for new technologies', *EPRI Journal* **7** (1): 6–13.

Linton, D. (1965) 'The geography of energy', *Geography* **L**: 197–228.

Lovins, A. B. (1978) 'Soft Energy technologies', *Annual Review of Energy* **3**: 477–517.

Luten, D. B. (1971) 'The economic geography of energy', *Scientific American*, 224, No. 3, 165–175.

Lysyk, K., E. H. Bohmer and W. L. Phelps (1977) *Alaska Highway Pipeline Inquiry*, Supply and Services Canada, Ottawa.

McCaslin, J. V. (ed.) (1982–87) *International Petroleum Encyclopedia*, Pennwell Publishing, Tulsa.

McDougall, G. H. G., J. R. B. Ritchie and J. D. Claxton (1979) *Energy Consumption and Conservation Patterns in Canadian Households – Executive Summary*, Mimeo, School of Business and Economics, Wilfrid Laurier University, Waterloo, Ontario.

McDougall, J. N. (1982) *Fuels and National Policy*, Butterworth, Toronto.

McGranahan, G., S. Chubb and R. Nathans (1979) *Patterns of Urban Household Energy Use in Developing Countries: The Case of Nairobi*, No. 116, Institute for Energy Research, State University of New York, Stony Brook, New York.

McGranahan, G. and M. Taylor (1977) *Urban Energy Use Patterns in Developing Countries: A Preliminary Study of Mexico City*, No. 113, Institute for Energy Research, State University of New York, Stony Brook, New York.

Magrath, C. A. (1919) *Final Report of Fuel Controller*, King's Printer, Ottawa.

Makhijani, A. and A. Poole (1975) *Energy and Agriculture in the Third World*, Ballinger, Cambridge, Mass.

Mankin, C. J. (1983) 'Unconventional sources of natural gas', *Annual Review of Energy* **8**: 27–43.

Manne, A. S. and L. Schrattenholzer (1984) 'International energy workshop: a summary of the 1983 poll responses', *The Energy Journal* **5** (1): 45–64.

Manners, G. (1964) *The Geography of Energy*, Hutchinson University Library, London.

Marshall, E. (1986) 'Fill the oil reserve academy report says', *Science* **232**: 441–2.

Maull, H. (1980) *Europe and World Energy*, Butterworth, London.

Maull, Hans W. (1984) *Raw Materials, Energy and Western Security*, Macmillan Press, London.

Maxey, M. N. (1980) 'Energy and ethical priorities', in *People Priorities: The Moral and Ethical Issues of Energy vs. Human Health and Well-Being*, Special Publication No. 18, IEEE Power Engineering Society, Piscataway, New Jersey.

Mitsch, W. J., R. K. Ragade, R. W. Bosserman and J. A. Dillon (1982) *Energetics and Systems*, Ann Arbor Science Publishers, Michigan.

Mohnfeld, J. H. (1984) 'The trend of structural change in the international oil industry in the 1980s', *Annual Review of Energy* **9**: 155–77.

Morgan, W. B., R. P. Moss and G. J. A. Ojo (eds) (1980) *Rural Energy Systems in the Humid Tropics*, The United Nations University, Tokyo.

Morrison, D. E. and D. G. Lodwick (1981) 'The social impacts of soft and hard energy systems', *Annual Review of Energy* **6**: 357–78.

Nader, L. and S. Beckeman (1978) 'Energy as it relates to the quality and style of life', *Annual Review of Energy* **3**: 1–28.

Nader, L. and N. Millerton (1979) 'Dimensions of the "people" problem in energy research and "the" factual basis of dispersed Energy futures', *Energy* **4**: 953–67.

Neff, T. L. (1984) *The International Uranium Market*, Ballinger, Cambridge, Mass.

Nelson, R. H. (1983) *The Making of Federal Coal Policy,* Duke University Press, Durham NC.

Newcombe, K., J. D. Kalma and A. R. Aston (1978) 'The metabolism of a city: the case of Hong Kong', *Ambio1* **7**: 3–15.

Nilsson, S. (1974) 'Energy analysis – a more sensitive instrument for determining the costs of goods and services', *Ambio* **III**: No. 6, 222–4.

Nuclear Energy Agency (NEA) (1978) *World Uranium Potential – An International Evaluation*, OECD, Paris.

Nuclear Energy Agency (1982) *Nuclear Energy and Its Fuel Cycle: Prospects to 2025*, OECD, Paris.

ODell, P. (1963) *An Economic Geography of Oil*, Bell, London.

ODell, P. (1969) *Natural Gas in Western Europe. A Case Study in the Economic Geography of Energy Resources*, Bohn, Haarlem.

ODell, P. (1970) *Oil and World Power: Background to the Oil Crisis*, Penguin, Harmondsworth.

ODell, P. (1974) *Energy: Needs and Resources*, Macmillan Press, London.

ODell, P. and K. E. Rosing (1983) *The Future of Oil, World Oil Resources and Use*, Kogan Page, London.

ODell, P. and K. E. Rosing (1985) East–West differ on estimates', *Petroleum Economist* **LII**: 9, 329–31.

Oil and Gas Journal (1986) 'USSR to pace hefty world gas supply; demand growth to 2000', **28** (38): 25–31.

Oilweek (1986a) *Canadian Pipeline Map 1985–86. Oilweek,* **37** (9): Supplement.

Oilweek (1986b) 'Refiners emphasize light products, operating efficiency', **37** (19): 11.

Oke, T. R. (1973) 'Urban heat island dynamics in Montreal and Vancouver', *Atmospheric Environment* **9**: 191–200.

References

Okrent, D. (1980) 'Comment on social risk', *Science* **208**: 372–5.
Organization for Economic Co-operation and Development (OECD) (1983) *Uranium Resources, Production and Demand*, OECD, Paris.
Osleeb, J.P. and I. M. Sheshkin (1977) 'Natural gas: a geographical perspective', *Geographical Review* **67**: 71–85.
Owen, A.R. (1985) *The Economics of Uranium*, Praeger, New York.
Pachauri, P. K. (1985) *The Political Economy of Global Energy*, The Johns Hopkins University Press, Baltimore.
Parker, L. B., R. L. Bamberger and S. R. Abbasi (1981) *The Unfolding of the Reagan Energy Program: The First Year*, Report No. 81–266 ENR, Congressional Research Service, Library of Congress, Washington DC.
Parker, S. P. (1981) *Encyclopedia of Energy*, McGraw-Hill, New York.
Parliament of Canada (1953) *Debates*, King's Printer, Ottawa.
Perry, A. M. (1978) *World Uranium Resources*, International Institute for Advanced Systems Analysis, Vienna.
Petroleum Economist (1986) 'Prospects for the natural gas trade', **LIII** (5): 158–9.
Procter, R. M., G. C. Taylor and J. A. Wade (1984) *Oil and Natural Gas Resources of Canada, 1983*, Paper 83-31, Geological Survey of Canada, Ottawa.
Puttagunta, V. R. (1975) *Temperature distribution of Energy consumed as heat in Canada*, Report No. 5235, Atomic Energy of Canada, Ottawa.
Quo, F. Q. (1986) 'Japan's resource diplomacy', Chapter 5 in Keith, R. C. (ed.), *Energy, Security and Economic Development in East Asia*, Croom Helm, London.
Regional Plan Association/Resources for the Future (1974) *Regional Energy Consumption*, Regional Plan Association, New York.
Revelle, R. (1976) 'Energy use in rural India', *Science* **192**: 969–75.
Roberts, F. S. and W. W. Waterman (1980) *Energy Modelling II. The Interface between Model Builder and Decision Maker*. Institute of Gas Technology, Chicago.
Rudman, R. L. and C. G. Whipple (1980) 'Time lag of energy innovation', EPRI Journal **5**: 3, 14–20.
Sagers, M.J. and A. Tretyakova (1986) 'Constraints on gas for oil substitution in the USSR: the oil industry and gas storage', *Soviet Economy* **2**: 1, 72–94.
Sassin, W. (1980) 'Urbanization and the energy problem', *Options* **3**: 1–4.
Sawyer, S. W. and J. R. Armstrong (eds) (1985) *State energy policy: current issues, future directions*, Westview Press, Boulder, Colorado.
Schanz, J. R. (1978) 'Oil and gas resources – welcome to uncertainty', *Resources*, No. 58, Resources for the Future, Washington DC.
Schipper, L., A. Keloff and A. Kahane (1985) 'Explaining residential energy use by international bottom-up comparisons', *Annual Review of Energy* **10**: 341–405.
Segal, J. (1980) 'Rapid growth in output expected', *Petroleum Economist* **xlviii**, No. 8, 336–7.
Sewell, W. R. D. (1964) 'The role of regional interties in post-war resource development', *Annals of the Association of American Geographers* **54** (4): 566–81.
Sewell, W. R. D. and H. D. Foster (1980) *Energy Conservation through Land Use Planning*, Working Paper No. 6, Lands Directorate, Environment Canada, Ottawa.
Shabad, T. (1984) 'News Notes (1983 performance and 1984 plans in the basic extractive and primary industries of the USSR)', *Soviet Geography*, **xxv** (4): 264–95.
Shabad, T. (1986) 'Geographic aspects of the new Soviet Five-Year Plan', *Soviet Geography* **xxvii** (1): 1–16.
Shwardran, B. (1986) *Middle East Oil Crisis Since 1973*, Westview Press, Boulder, Colorado.
Simpson, R. A., D. M. Nowland and D. W. Rutledge (1961) *The Natural Gas Industry in Canada, 1960*, Department of Mines and Technical Surveys, Ottawa.
Simpson, R. A. and D. W. Rutledge (1964) *The Natural Gas Industry in Canada: 1961*

and 1962, Mineral Information Bulletin, MR 72, Department of Mines and Technical Surveys, Ottawa.

Slesser, M. (1978) *Energy in the Economy*, Macmillan Press, London.

Smit, B. and T. Johnston (1983) 'Public policy assessment: evaluating objectives of resource policy', *Professonal Geographer* **35**: 172–8.

Statistics Canada (1978) *Human Activity and the Environment*, 11-509E, Occasional, Supply and Services, Ottawa.

Statistics Canada (1931–1987) *Crude Petroleum and Natural Gas Production*, No. 26-006, Supply and Services, Ottawa.

Statistics Canada (1926–1987) *The Crude Petroleum and Natural Gas Industry*, No. 26-213, Supply and Services, Ottawa.

Statistics Canada (1921–1987) *Coal and Coke Statistics*, No. 45-002, Supply and Services, Ottawa.

Statistics Canada (1986) *Quarterly report on energy supply–demand in Canada*, No. 57-003, Supply and Services, Ottawa.

Stern, P. C. (1984) *Energy Use: The Human Dimension*, W. H. Freeman, New York.

Steward, F. R. (1978) 'Energy consumption in Canada since confederation', *Energy Policy* **6** (3): 239–45.

Summers, C. M. (1971) 'The Conversion of Energy', *Scientific American* **224**, 134–63.

Taher, A. H. (1980) *Energy: A Global Outlook. The Case for Effective International Co-operation*, Pergamon Press, Oxford.

Taylor, P. B. (1982) *The Iron and Steel Industry*, Report No. 16, Energy Audit Series, UK, Department of Energy and Department of Industry, London.

Thompson, A. R. (1978) *West Coast Oil Ports Inquiry: Statement of Proceedings*, West Coast Oil Ports Inquiry, Vancouver BC.

Thornthwaite, C. W. (1961) 'The task ahead', *Annals of the Association of American Geographers* **51**: 345–56.

Thornton, J. (1986) 'Chernobyl and Soviet Energy', *Problems of Communism*, **xxxv**, Nov–Dec, 1–16.

Tribus, M. and E. C. McIrvine (1971) 'Energy and information', *Scientific American* **224** (3): 179–88.

UK, Department of Energy (DOE) (1976) *Passenger Transport: Short and Medium Term Considerations*, Energy Paper No. 10, HMSO, London.

UK, Department of Energy (1977a) *Report of the Working Group on Energy Elasticities*, Energy Paper No. 17, HMSO, London.

UK, Department of Energy (1977b) *Freight Transport: Short and Medium Term Considerations*, Energy Paper No. 24, HMSO, London.

UK, Department of Energy (1978) *Energy forecasting methodology*, Energy Paper No. 29, HMSO, London.

UK, Department of Industry (1979) *The Aluminum Industry*, Report No. 6, Energy Audit Series, Department of Industry, London.

United Nations, Department of Economic Affairs (1952) *World Energy Supplies in Selected Years, 1929–1950*, Statistical Papers, Series J., No. 1, UN, New York.

United Nations, Department of Economic Affairs (1957) *World Energy Supplies, 1951–1954*. Statistical Papers, Series J, No. 2, UN, New York.

United Nations, Department of Economic and Social Affairs (1965) *World Energy Supplies, 1960–1963*, Statistical Papers, Series J, No. 8, UN, New York.

United Nations, Department of Economic and Social Affairs (1976) *World Energy Supplies*, Statistical Papers, Series J, UN, New York.

United Nations, Department of International Economic and Social Affairs (1981) *Yearbook of World Energy Statistics*, UN, New York.

United Nations, Department of International Economic and Social Affairs (1985) *Energy Balances and Electricity Profiles, 1982*, UN, New York.

References

United Nations, Department of International Economic and Social Affairs (1986) *Energy Statistics Yearbook, 1984*, UN, New York.

United Nations, Department of International Economic and Social Affairs (1987) *1985 Energy Statistics Yearbook*, UN, New York.

US, Department of Energy (DOE) (1985) *The National Energy Policy Plan*, US Department of Energy, Washington DC.

US, Department of Energy (1987) *State Energy Data Report: Consumption Estimates, 1960–1985*, Energy Information Administration, Washington DC.

US, Geological Survey (1975) *Geological Estimates of Undiscovered Recoverable Oil and Gas Resources in the United States*, Geological Survey Circular 725, USGS, Washington DC.

US, Office of Technology Assessment (1982) *Energy Efficiency of Buildings in Cities*, Superintendent of Documents, Washington DC.

Urquhart, M. C. (1965) *Historical Statistics of Canada*, Macmillan, Toronto.

Wagstaff, H. R. (1974) *A Geography of Energy*, W. C. Brown, Dubuque, Iowa.

Ward, P. C. (1974) 'The implications of national policies on world energy', Chapter 8 in M. S. Macrakis (ed.) *Energy: Demand, Conservation and Institutional Problems*, MIT Press, Cambridge, Mass.

Wasp, E. J. (1983) 'Slurry pipelines', *Scientific American* **249** (5): 48–55.

Watkins, G. C. and M. A. Stabback (1981) *Reaction: the National Energy Program*, Fraser Institute, Vancouver BC.

Weinberg, A. M. (1966) 'Can technology replace social engineering?' *Bulletin of Atomic Scientists* **22** (10): 4–8.

Weinberg, A. M. (1980) 'Limits to Energy modelling', in F. S. Roberts and W. W. Waterman, *Energy Modelling II: The Interface between Model Builder and Decision Maker*, Institute of Gas Technology, Chicago, 23–33.

Weinberg, A.M. (1986) 'Are breeder reactors still necessary?', *Science* **232**: 695–6.

Wilbanks, T. J. (1985) 'Geography and energy: the quest for roles and missions', Chapter 25 in F. J. Calzonetti and Solomon, *Geographical Dimensions of Energy*, D. Reidel Publishing Company, Dordrecht.

Wilson, C. L. (1980) *Coal: Bridge to the Future*, Ballinger, Cambridge, Mass.

Wilson, D. (1986) 'Serious implications in oil shortfall in the USSR', *Petroleum Review* **40** (468): 18–19.

Winner, L. (1982) 'Energy regimes and the ideology of efficiency', Chapter 13 in G. H. Daniels and H. Rose (eds) *Energy and Transport: Historical Perspectives on Policy Issues*, Sage Publications, Beverley Hills, California.

Wood, T. S. and S. Baldwin (1985) 'Fuelwood and charcoal use in developing countries', *Annual Review of Energy* **10**: 407–29.

Workshop on Alternative Energy Strategies (WAES) (1977) *Energy: Global Prospects, 1985–2000*, McGraw-Hill, New York.

World Bank (1983) *The Energy Transition in Developing Countries*, World Bank, Washington, DC.

World Energy Conference (1978) *World Energy: Looking Ahead to 2020*, IPC Science and Technology Press, London.

World Energy Conference (1983a) *Oil Substitution: World Outlook to 2020*, Graham and Trotman, London.

World Energy Conference (1983b) *Energy 2000–2020: World Prospects and Regional Stresses*, Graham and Trotman, London.

Zimmermann, E. W. (1950) *World Resources and Industries: A functional appraisal of the availability of agricultural and industrial materials*, Harper, New York.

Index

Index

Index

Index

Index